投考公務員系列

入境事務處
筆試全攻略

Fong Sir 著

目錄

CHAPTER

1

LANGUAGE
& APTITUDE
TEST

A. Comprehension

【定義】

Comprehension（或稱Reading Comprehension，中文叫「閱讀理解」）測試，旨在測試考生理解書面文本的能力，文本的內容通常為非技術性材料。

【技巧】

1. 兩種方法

a. **「先讀內文，後答題目」**：適合簡單易懂的文章，且時間較充裕的情況。

b. **「先看題目，後讀內文，再答問題」**：這種方法在文章較長的情況下合用。考生可以帶著問題閱讀：當內文與考題有關時，細讀；與題目無關時，跳讀。

2. 三個側重點

閱讀時要注意避免逐字逐句細讀，而要有所側重。側重點包括：

a. **「首尾段」和「首尾句」**：這些地方往往是作者所要談及的論點、主題或中心思想。

b. 語篇標誌詞：把握住這些詞語，就等於把握了句與句、段與段之間的關係。語篇標誌詞主要表示：例子、列舉、比較、結論、原因、結果、目的、時間、地點、方式手段、分類、補充、強調、條件、轉折和對比。

c. 長句、難句：事實上，不少問題就是針對這些長句、難句而設。

3. 題型

a. 主旨題：旨在測試考生把握主題與中心思想的能力，主要形式有：

形式	例句
(1) 問中心思想	The main idea of this passage is...
	The passage mainly discusses...
	What is the passage primarily concerned about?
(2) 問寫作目的	The author writes this passage mainly to...
	The author's purpose in writing this passage is...
(3) 問態度	The author's attitude towards... is...
	The tone of this passage can be described as...

b. 細節題：這類問題測試考生把握文章細節的能力，主要與文中的考點相關聯，如：最高級、唯一性、其他對比、數字年代、原因等。主要形式有：

形式	例句
(1) 是非題	Which of the following is True?
	Which of the following is NOT True?
	Which of the following is NOT mentioned in paragraph...?
(2) 例證題	The author gives an example in paragraph... mainly to show that...
(3) 其他具體細節題	From the passage, it can be seen that...
	The main reason for... is...

c. 推理題：主要測試考生能否在理解字面意義的基礎上，根據所讀材料進行判斷和推論，進而理解文章的深層意思。主要形式有：

(1) It is implied in the passage that...

(2) The passage implies (suggests) that...

(3) It can be inferred from the passage that...

d. 詞義題：這類問題主要測試考生使用詞語搭配，以及根據上下文判斷詞義的能力。主要形式有：

(1) The word "..." in line (Paragragh) ... most probably means...

(2) In paragraph... , the word "..." refers to (stands for)...

(3) The word "..." in Paragraph... can be best replaced by...

4. 猜字的幾種技巧

在閱讀過程中，經常會遇到一些生詞、難詞，這時考生就需要利用猜字的技巧了，包括：

a. 利用詞根、詞綴構詞法

b. 直接定義：作者在行文中有時不得不使用某些難詞、偏詞，為使讀者容易理解，作者常常會在文章中直接解釋該詞語，考生就可以將之看作「提示字眼」，這些提示包括：(1) that is (to say); (2) e.g.; (3) or, in other words; (4) to put it in another way等。

例句：She is bilingual. In other words, she speaks English and French equally well.（註：bilingual是指「懂得說兩種語言」，與「English and French」對應）

c. 近義複述：同一短文中前後兩個句子、短語或單詞通常有相互解釋的作用，考生可從上下文的複述中，獲取與某一單詞或短語相關的信息以猜測詞義。

例句：It is difficult to list all of my father's attributes because he has so many different talents and abilities. (註：attributes的意思為「特質和才能」，與句子後段的「many different talents and abilities」互相對應)

d. **對比和並列表述**：利用上下文的對比或並列，表達猜測詞義是最常用、最可靠的方法之一。有不少句子會在上下文中給出某個生詞的同義或反義詞，運用對比或並列表達對這些生詞加以提示。通過了解詞與詞之間的連接關係，特別是一些語篇標誌詞，如：(1) however; (2) on the other hand; (3) nevertheless等，考生便可推斷這些生詞的詞義。

例句：If you agree, write "yes"; if you dissent（不同意），write "no".

e. **根據常識**：有些生詞看似很難，但根據語境、根據讀者的經歷或常識，就很易猜出詞義。如所讀的材料是考生熟悉的內容，或在自己專業知識的範圍內，生詞就更易化解了。

【練習】

Read the following passage and answer questions below. For each question, choose the best answer from the given choices.

Political education has many connotations. It may be defined as the preparation of a citizen to take well informed, responsible and sustained action for participation in the national struggle in order to achieve the socio-economic objectives of the country. The predominant socio-economic objectives in India are the abolition of poverty and the creation of a modern democratic, secular and socialist society in place of the present traditional, feudal, hierarchical and in egalitarian one.

Under the colonial rule, the Congress leaders argued that political education was an important part of education and refused to accept the official view that education and politics should not be mixed with one another. But when they came to power in 1947 they almost adopted the British policy and began to talk of education being defiled by politics. "Hands off education" was the call to political parties. But in spite of it, political infiltration into the educational system has greatly increased in the sense that different political parties vie with each other to capture the mind of teachers and students. The wise academicians wanted political support, without political interference. What we have actually received is infinite political interference with little genuine political support. This interference with the educational system by political parties for their own ulterior motives is no political education at all and with the all round growth of elitism, it is hardly a matter for surprise that real political education within the school system (which really means the creation of a commitment to social transformation) has been even weaker than in the pre-independence period.

During that time only, the struggle for freedom came to an end and the major non-formal agency of political education disappeared. The press played a major role by providing some political education. But it did not utilize the opportunity to the full and the strangle hold of vested interests continued to dominate it. The same can be said of political parties as well as of other institutions and agencies outside the school system which can be expected to provide political education. After analyzing all these things, it appears that we have made no progress in genuine political education in the post-education period and have even slided back in some respects. For instance, the education system has become even more elite-oriented. Patriotism has become the first casualty. The father of the nation gave us the courage to oppose government when it was wrong, in a disciplined fashion and on basic principles. Today, we have even lost the courage to fight on basic issues in a disciplined manner because agitational and anarchic politics for individual, group or party aggrandizement has become common. In the recent times the education system continues to support domination of the privileged groups and domestication of the under- privileged ones. The situation will not change unless we take vigorous steps to provide genuine political education on an adequate scale. This is one of the major educational reforms we need, and if it is not carried out, mere linear expansion of the existing system of formal education will only support the status quo and hamper radical social transformation.

1. Which word is nearly opposite in meaning as “defile” as used in the passage?

   A. disparage
   B. forgery
   C. degenerate
   D. sanctify

2. According to the passage, what should be the main purpose of political education?

   A. To champion the cause of elitism
   B. To bring qualitative change in the entire education system
   C. To create an egalitarian society
   D. To prepare the young generation with high intellectual acumen

3. How has politics been related to educational institutions after independence?

   A. Although they got political support but there was no interference of politics
   B. It is clear that they got almost no political support as well as political interference
   C. They got political support at the cost of political interference.
   D. There was substantial interference without political support

Answer:

1. **D** The word "defile" means to make foul, dirty, or unclean. Disparage means to speak of or treat slightingly. Forgery means the act of reproducing something for a deceitful or fraudulent purpose. Sanctify means to purify or free from sin.

2. **C** The answer to this question is given in the 1st paragraph. It defines the purpose of political education: it is to guide the citizens to work for the socio economic objectives of the country. And according to this paragraph the socio economic objectives of India are the abolition of poverty and the creation of a modern democratic, secular and socialist society in place of the present traditional, feudal, hierarchical and in egalitarian one in short an egalitarian society.

3. **D** It is mentioned in the first paragraph 9th line: "What we have actually received is infinite political interference with little genuine political support."

B. Cloze

【定義】

Cloze（或稱Cloze Test，即「填空題」）指在閱讀理解能力卷中的填充題，通常提供某段原文，要求考生在空白的橫線上選擇合適的文字，令文句的意思變得通順、合理。

【技巧】

1. 運用語篇知識

a. 從段首句或首段找答案

通常，填空題的首句（甚至前幾句話）都是完整的，為考生提供足夠訊息解構文章思路，而這裡往往包含主題句，或為理解文章的大意和主要內容提供必要的線索。

b. 利用定義句解題

考生在閱讀中如遇到不理解的生詞或關鍵詞，應從短文的上下文中找出能夠為其定義的短語或句子。在尋找定義線索時，is、mean、is called等詞語可成為語言暗示。有時，反義詞語也能為推斷生詞詞義提供幫助。

c. 把握文章發展的基本線索

填空題是一篇有完整內容且按照一定思路發展的文章，各段落與各句之間都有邏輯上的必然關係：論說文一般按照邏輯推理關係展開；記敘文往往按照時間順序來鋪陳；描寫文的發展常常表現為空間關係。

答題過程中，考生首先要樹立起語篇概念，抓住文章主題思想，釐清其結構布局，明辨句與句之間、段與段之間的關係，不要只拘泥於句子、語法。利用句與句，句群與句群之間的邏輯關係解題。

文章的邏輯關係主要包括列舉、原因、結果、讓步、對照、補充、時間順序、目的、條件等，而這些邏輯關係又是靠邏輯詞來表達的。沒有邏輯詞，文章就顯得語義模糊不清，不能形成篇章。

考生應熟記用來表達不同邏輯關係的連詞（見附表）：

連詞表達的意思	用字
(1) 列舉	first, second, third...; firstly, secondly, thirdly..., first, next, then...; in the first place, in the second place...; for one thing, for another thing...; to begin with, to conclude...
(2) 原因	because, since, as, now that...
(3) 結果	so, therefore, thus, hence, accordingly, consequently, as a result...
(4) 讓步、轉折	however, nevertheless, nonetheless, still, though, yet, in spite of, at any rate, in any case, whoever, whatever...
(5) 對照	on the contrary, in contrast, by contrast, in comparison, by comparison, conversely...
(6) 補充	also, further, furthermore, likewise, similarly, moreover, in addition, what is more, too, either, neither, not...but..., not only...but also...
(7) 時間順序	when, while, as, after, before, since, until, as soon as, once...
(8) 目的	that, so that, lest, for fear that...
(9) 條件	if, suppose, unless, in case, so (as) long as, so far as, on condition (that), provided (that)...

d. 利用上下文尋找解題信息

填空題的文章都是一個意義相關聯的語篇。它圍繞一個話題論述，在行文中詞語重複，替代現象是不可避免的。所以在解題時，考生應從審視上下文尋找相關線索，有時只需將文中的詞或短語照搬即

可。如果上下文的線索以語義照應的形式出現，考生可利用推斷方法將相關語義連接起來。

2. 運用背景知識和社會常識解題

由於填空題的文章的內容多與日常生活相關，有的內容是我們頭腦中已經了解的常識。當對語言的把握不很準確時，可充分利用社會知識和科普常識來幫助判斷，先找出並理解文章主題、主線；並根據主題猜測細節；注意從重複出現的詞語中尋找，體會文章表達的氛圍。

3. 運用詞彙、語法等知識

a. 充分發揮平時積累的語言基礎知識，利用搭配知識解題

注意文中的邏輯搭配（包括過渡詞、連接手段、指代關係、肯定、否定等）；語義搭配（包括區別同義詞、近義詞、反義詞、形近異義詞、同形異義詞）；結構搭配（指名詞、動詞、形容詞等在句中或文中與其他詞的搭配要求）；慣用搭配（即通常所說的固定短語）。

b. 利用詞根詞綴的知識解題

要牢記常用的詞根（root）和詞綴（affix），特別是構詞能力強的詞根。記憶詞根詞綴時，要利用已經熟知的詞的詞根詞綴去加強對詞根詞綴的記憶，然後將詞根詞綴的知識運用到生詞中去。平時閱讀時，應學會辨認、分析、推斷同根詞的意義。遇到長詞時，應充分利用詞根詞綴的知識將單詞進行分解，找出其意義的根據。詞根詞綴知識對解答完形填空中的詞彙類題最有實踐意義。

c. 進行語法分析，縮小選擇範圍

分析題區的句法關係，是簡單句、並列句，還是複合句；判斷所填內容在句中充當什麼成分，應是什麼詞性。另外，從時態、語態、語氣、名詞的數等各個角度分析所填內容是否與上下文一致，從而縮小選擇範圍。

4. 絕對不能放棄利用以上方法仍解不出來的題目

可利用平時培養的語感，將選項放入題區，反覆誦讀，哪個上口就選哪個。如果仍不能選出，就選一個在所有選項中出現頻率較低的選項。

【練習】

Read the following passage and answer questions. For each question, choose the best answer from the given choices.

Before the 20th century the horse provided day to day transportation in the United States. Trains were used only for long-distance transportation.

Today the car is the most popolar ___(1)___ of transportation in all of the United States. It has completely ___(2)___ the horse as a means of everyday transportation. Americans use their car for ___(3)___ 90 percent of all personal ___(4)___ .

Most Americans are able to ___(5)___ cars.The average price of a ___(6)___ made car was, 1,050 in 1950, 1,740 in 1960 and up to 2,750 ___(7)___ 1975. During this period American car manufacturers set about ___(8)___ their products and work efficiency.

As a result, the yearly income of the ___(9)___ family increased from 1950 to 1975 ___(10)___ than the price of cars. For this reason ___(11)___ a new car takes a smaller ___(12)___ of a family's total earnings today.

In 1951 ___(13)___ it took 8.1 months of an average family's ___ (14)___ to buy a new car. In 1962 a new car ___(15)___ 8.3 of a family's annual earnings. By 1975 it only took 4.75 ___(16)___ income. In addition, the 1975 cars were technically ___(17)___ to models from previous years.

The ___(18)___ of the automobile extends throughout the economy ___(19)___ the car is so important to Americans. Americans spend more money to ___(20)___ their cars running than on any other item.

(1) A. kinds   B. means   C. mean   D. types

(2) A. denied   B. reproduced   C. replaced   D. ridiculed

(3) A. hardly   B. nearly   C. certainly   D. somehow

(4) A. trip   B. works   C. business   D. travel

(5) A. buy   B. sell   C. race   D. see

(6) A. quickly   B. regularly   C. rapidly   D. recently

(7) A. on   B. in   C. behind   D. about

(8) A. raising   B. making   C. reducing   D. improving

(9) A. unusual   B. interested   C. average   D. biggest

(10) A. slowest   B. equal   C. faster   D. less than

(11) A. bringing   B. obtain   C. bought   D. purchasing

(12) A. part   B. half   C. number   D. side

(13) A. clearly   B. proportionally   C. percentage   D. suddenly

(14) A. income   B. work   C. plants   D. debts

(15) A. used   B. spend   C. cost   D. needed

(16) A. months   B. dollars   C. family   D. year

(17) A. famous   B. superior   C. fastest   D. purchasing

(18) A. running   B. notice   C. influence   D. discussion

(19) A. then   B. as   C. so   D. which

(20) A. start   B. leave   C. keep   D. repair

答案：

1. **B**　本題易錯選C，但mean不是名詞，「手段」和「方式」的名詞為means。

2. **C**　根據句子意思判斷，replace意為「代替」，正確。

3. **B**　Nearly的意思是「幾乎，大約」；hardly的意思是「幾乎不」；如：He could hardly do that。（他幾乎不能做那件事）

4. **A**　本題易錯選D，但travel指遠距離的旅行，而trip指以工作和娛樂為目的的短距離旅行。

5. **A**　根據下句的意思判斷。

6. **D**　根據句子意思判斷，recently made意為「最近生產的」。

7. **B**　在某一年用介詞in。

8. **D**　因為「改進產品」與「提高工作效率」意思連貫，符合上下文意思。

9. **C**　句子意思為「平均家庭年收入」，所以應選average。

10. **C**　「than」的前面要用比較級。

11. **D**　根據句子意思判斷，purchasing為交易買賣的意思。

12. **A**　take a part在本句中意為「佔一部分」。

13. **B**　這裡需要一個副詞proportionally表達「按比例地、適當地、相稱的、相當的」意思。

14. **A**　Income在上文已提及。

15. **C**　Cost指某物花費某人多少錢，如：The coat cost me。「Spend」指某人花多少錢買某物，如：I spent on the coat.

16. A 根據句子意思判斷。

17. B 「superior to」是短語，有「優於」的意思。

18. C 根據句子意思判斷。

19. B as引導的原因狀語從句。

20. C 「keep their car running」是屬於keep something doing 這個固定動詞短語，意為「保持……繼續進行」。

C. Error Identification

【定義】

Error Identification（又稱Error Recognition，中文叫「改錯題」）是指每題一個句子，句子中會有四個地方畫上底線，考生必須從中揀出一個文法錯誤的項目。

【技巧】

1. 逐字逐句閱讀。不要只查看句子的下劃線部分，因為錯誤通常只是因為句子的上下文不符而不正確。

2. 不要讀得太快。否則將有可能會跳過錯誤，特別是那些涉及「小詞」（介詞、代詞、文章）的錯處。

3. 嘗試在腦中「唸出」每個單詞，這將助你捕獲「聽」起來錯誤的地方。

4. 當閱讀後無法找到錯誤，請查看句子中的動詞以查看它們是否正確使用，檢查的地方包括：時態，形式及與主題的一致性。

5. 如果動詞似乎使用正確，請檢查：單詞選擇、單詞形式、介詞使用等。

6. 如果仍然找不到錯誤，請刪除看似正確的選項。如果仍有多個選擇，猜吧！在您不確定的項目旁邊的答題紙上留下標記。（但要確保在測試結束前擦掉）

7. 永遠不要在某條題目花上太多時間

【練習】

Each of the sentence below may contain a language error. Identify the part (underlined and lettered) that contains the error or choose "(E) No error" where the sentence does not contain an error.

1. The minimum wage <u>bill</u> provides for a gradual <u>reduction</u> of the <u>minimum</u> wage from $5.15 an hour to $7.25 an hour <u>over two years</u>.

   A. bill
   B. reduction
   C. minimum
   D. over two years
   E. no error

2. Advocacy for child war victims, children <u>in</u> hazardous work, abused children and <u>those</u> variously exploited or handicapped <u>has attracted</u> the attention and commitment of legislators and policy-makers <u>through</u> the world.

   A. in
   B. those
   C. has attracted
   D. through
   E. no error

3. Education <u>should</u> emphasize <u>our</u> interdependence <u>with</u> peoples, with other species and with the planet <u>as a whole</u>.

   A. should
   B. our
   C. with
   D. as a whole
   E. no error

4. <u>Most of</u> the large industries <u>in the</u> country <u>are</u> well organised and structured and are sometimes <u>backed up</u> internationally reputable mother companies.

   A. most of
   B. in the
   C. are
   D. backed up
   E. no error

5. <u>Should we</u> really <u>speak of</u> the "breakdown" of families <u>when we are</u> perhaps witnessing new family forms and a new social structure <u>arising</u> late capitalism?

   A. should we
   B. speak of
   C. when we are
   D. arising
   E. no error

6. The economy is heavy dependent on industry, and economic growth has always been of greater concern than environmental preservation.

   A. heavy
   B. on industry
   C. has always been
   D. than
   E. no error

7. The increase population, and rapid economic growth in recent years, have put a large and increasing stress on the water resources and environment in Ho Chi Minh City, Vietnam.

   A. increase
   B. in recent years
   C. on the water resources
   D. in
   E. no error

Answer:

1. **B** It should be "increase", not "reduction".
2. **D** It should be "throughout".
3. **C** It should be "with other peoples".
4. **D** It should be "backed up" by internationally reputable mother companies.
5. **D** It should be "arising from"
6. **A** It should be "heavily dependant"
7. **A** It should be "increase in population"

D. 閱讀理解

【定義】

閱讀理解著重測試考生對文段的整體把握力，考生需要閱讀的材料多為議論文，有些論點可以在文章中直接體現，但有些卻需要通過一定的推理才能得出，但材料中一般已包含論據和論證的過程。

【技巧】

我們根據選項、材料內容、材料內部邏輯關係的不同特點，歸納出幾種較常用的解題方法：

1. 選項排除法

【定義】解題時可直接從對比選項入手，快速排除明顯不正確的選項：

a. 當既有指出問題的選項，又有給出對策的選項時，通常給出對策的選項為正確答案。

b. 正確答案給出的對策應具有積極的一面，且切實可行。

c. 明顯違背常識的選項可直接排除。

【例題】哲學曾經是一種生活方式。所謂蘇格拉底的哲學，不只是他和別人對話的方法，以及他在對話中提出的種種理論，更是他不立文字、浪跡街頭、四處與人閒聊的生活方式。哲學從一開始就不是一種書面的研究，而是一種過日子的辦法。只不過我們後來都忘了這點，把它變成遠離日常的艱深遊戲。即使是很多人眼中刻板的康德，也不忘區分「學院意義的哲學」和「入世意義的哲學」，並以後者為尊。

以上的文字意在說明：

A. 哲學源於生活應服務於人民

B. 如今的哲學發展偏離了它的本質

C. 康德和蘇格拉底的哲學觀念一脈相承

D. 當代人們對哲學的詮釋方式發生了改變

【答案】A。選項B至D都只是提出一個問題或現象，而A項中「應服務於人民」提出了對策，具明顯傾向性，表明作者的態度，且不違背常理，由此正確答案為A。

2. 直接摘取法

【定義】某些題目的主旨在材料中已有明確表述,考生只需把它識別出來,找出與之對應的選項即可。此類材料一般不僅指出問題,更提供了解決方法。

【例題】從古至今都沒有長生不老的人,但每個人的壽命都極不相同,從生物學和醫學上來看,人類的壽命應該有一個生物學上的最大值,不過目前尚不能確定人類的最高壽命有多高。

上述主要支持一個觀點,即人的壽命:

A. 是無限的

B. 究竟有多高,我們是無法知道的

C. 應該有極限,但目前還未被發現

D. 對不同的人是不一樣的

【答案】C。注意文段中的關鍵詞「應該……」,它所在的句子「人類的壽命應該有一個生物學上的最大值,不過目前尚不能確定人類的最高壽命有多高」主要包含了兩個信息:一是人的壽命有極限,二是這個極限目前還不能確定。包含這兩點的只有C。

3. 對症下藥法

【定義】與「直接摘取法」相對應的是「對症下藥法」，適用於此類方法的文段材料一般只提出了問題，沒有給出解決辦法，而這個解決辦法便是它的主旨。

【例題】作為一個擁有五千年不間斷文明史的古國，中國擁有十分豐富的非物質文化遺產。這些活態的文化不僅構成了中華民族深厚的文化底蘊，也承載著中華民族文化淵源的基因。但隨著中國現代化建設的加速、文化標準化以及環境條件的變化，尚有不計其數的文化遺產正處於瀕危狀態，它們猶如一個個影子，隨時都可能消亡。

對這段文字概括最準確的是：

A. 文化遺產保護工作要有新思路

B. 要重視現代化建設帶來的新問題

C. 新形勢下亟需加強文化遺產保護

D. 諸多因素威脅著文化遺產的生存狀態

【答案】C。作者指出的問題是眾多文化遺產正處於瀕危狀態。針對這一情況,其對策顯然是要加強保護,而且「亟需」也恰當地體現出了保護的迫切性,故C正確。

4. 概念提示法

【定義】閱讀理解的文段常會在結尾引入一個具有特定意義的概念,這一概念就是文段的落腳點,文段的主旨一般都與這個概念有關。考生只要抓住這個概念,再運用排除法,通常可以快速鎖定答案。

【例題】中國古代禮制要求服裝盡力遮掩身體的各種凹凸,在裁製冕服時可以忽略人體各個部位的三維數據,不需要進行細緻的測量。冕服章紋要有效地體現等級區別,圖案就必須清晰可辨、鮮明突出,這使中國古代服飾中與服飾圖案相關的繪、染、織、繡等工藝技術相當發達,也使中國古代服裝的裁製向著有利於突出圖案的方向發展。與西方重視身體三維數據、要求服裝緊窄合體的立體剪裁法不同,中國古代無論是冕服對人所佔空間的擴大,還是圖案對冕服平面風格的要求,都指向了中國傳統服裝寬大適體的平面剪裁法。

這段文字意在說明：

A. 禮制的要求使中國傳統服裝採用了平面剪裁法

B. 中國古代服裝的剪裁方法推動了印染技術發展

C. 中西方剪裁方法的分化以冕服的產生與發展為特徵

D. 禮制對官員服裝的規定制約了中國古代服飾藝術的發展

【答案】A。文段的末尾出現了一個專業概念「平面剪裁法」，主旨應包含這一概念，符合此要求的只有A。

5. 歸納論證法

【定義】有些閱讀理解的文段列舉了很多事例或包含很多方面的內容，解答此類題目時，需要運用歸納論證法來提煉文段的論點。歸納論證是一種由個別到一般的論證方法，它通過許多個別的事例或分論點，歸納出事物所共有的特徵，從而得出一個一般性的結論。

【例題】在安科萊，以畜牧為生的希馬人和以農業為生的伊魯人共同居住；在亞利桑那，納瓦霍人以前靠狩獵和採集為生、現在主要以畜牧為生，他們與經營農業的霍皮人為鄰；澳洲東南沿海地帶以

前住著以漁獵和劫掠糧食為生的土著居民,現在卻住著從事農業、畜牧業及工業的歐洲人。

作者列舉這些事實意在說明:

A. 環境迫使人們接受某生活方式

B. 人們對自然環境有很強的適應能力

C. 不同文化的族群完全有可能和諧相處

D. 地理環境並非人類生產方式的決定因素

【答案】D。文段列舉的這些事例看似很散亂,其實通過簡單地歸納即可得出:以畜牧為生、以農業為生、靠狩獵和採集為生、以漁獵和劫掠糧食為生、從事農業、畜牧業及工業等都是當地人謀生的手段,即他們的生產方式。由此可初步確定答案為D。

另外,居住在同一個地理區域的人類,不僅與同時代鄰近族群的生產方式不同,而且在以前與現在的縱向生產方式對比上,也是迥異的。由此可知,「地理環境並非人類生產方式的決定因素」。

6. 演繹論證法

【定義】與歸納論證法相反，「演繹論證法」就是由普通性前提，推出特殊性結論。

【例題】公共領域的問題從來都不是科學的問題、統計數字的問題，而是主觀認知的問題、意見的問題、想像的問題。在社會領域，人民的主觀感覺才是最主要的客觀事實。因此，在討論公共政策問題的時候，看似準確的數字，其實並不像官員、專家們想像的那樣重要。相反，如果一個國家的人民感覺自己的稅負沉重，那麼，該國的稅負就是沉重的，不論專家們計算出來的宏觀稅負水平與其他國家相比有多低。

這段文字旨在告訴我們：

A. 有時公共政策的科學制訂，需要我們遠離數據分析。

B. 一個國家稅負的沉重與否，取決於該國公民對稅負的主觀感受。

C. 我們對一些公共領域問題的認識有失偏頗

D. 解決公共領域問題時，參考人民的感覺比數據更重要。

【答案】B。文段採用的是演繹論證的方法：作者先提出一個普遍的規律，即在討論公共政策問題時，人民的感覺才是最重要的，進而由一般到特殊，推及到公共領域中的稅負問題，說明稅負是否沉重取決於該國人民對稅負的主觀感受。故B項符合題意，其他三項未體現稅務負擔這一論述主題，排除。

7. 因果論證法

【定義】因果論證是議論文最常採用的論證方法。一般使用了因果論證法的題幹，文段中會出現表因果邏輯的指示詞，考生根據指示詞，可判斷文段的主旨。一般來說，若文段中出現類似「因為……所以……」、「因……便……」、「由於……因而……」、「由於……因此……」的話，那麼後面的內容便是文段的主旨。

注意：若出現類似「……難怪……」的句式，則「難怪」前面的內容就是核心內容。

【例題】「官員」這種人從早到晚都在解決有答案的問題，頭腦已經被訓練成這樣，因此不是當政治人物的料。政治人物的任務是處理沒有答案的問題，去應付很有可能無解的問題。然而，一個人越是被訓練成優異的官員，就越以為問題都是有答案的，一定可以解

決，因此一遇到可能無解的問題就束手無策，甚至拍拍屁股一走了之。所以，一個人越是被訓練成官員，就越會變成不稱職的政治人物。

對這段文字的主旨概括最準確的一項是：

A. 一個優異的官員總以為問題都是有答案的

B. 政治人物的任務是處理沒有答案的問題

C. 不是所有的官員都能成為政治人物

D. 最優異的官員是差勁的政治人物

【答案】D。先將文段以句號劃分，分成以下四部分：

（i）「官員」這種人從早到晚都在解決有答案的問題，頭腦已經被訓練成這樣，因此不是當政治人物的料。

（ii）政治人物的任務是處理沒有答案的問題，去應付很有可能無解的問題。

（iii）然而，一個人越是被訓練成優異的官員，就越以為問題都是有答案的，一定可以解決，因此一遇到可能無解的問題就束手無策，甚至拍拍屁股一走了之。

（iv）所以，當一個人越是被訓練成官員，就越會變成不稱職的政治人物。

在（i）和（iii）中，都出現了表示因果關係的「因此」，而（iv）有「所以」。對文段進行層次劃分可知，由（i）至（iii）這三句話，與（iv）存在因果關係，「所以」後的內容為主旨，即D為答案。

8. 假設推斷法

【定義】當文段出現類似「如果……就……」的句式，通常強調的是「如果」等用字後面假設的條件，與此相反的做法是作者提倡的，即文段的主旨：「即使……也……」類，屬於讓步性假設關係，通常強調的是「也」等後面表示轉折的內容。

【例題】政府每提出一項經濟政策，都會改變某些利益集團的預期。出於自利，這些利益集團總會試圖通過各種行為選擇，來抵消政策對他們造成的損失，此時如果政府果真因此而改變原有的政策，其結果不僅使政府提出的政策失效，更嚴重的是使政府的經濟調控能力因喪失公信力而不斷下降。

這段文字主要論述了：

A. 政府制訂經濟政策遇到的阻力

B. 政府要對其制訂的政策持續貫徹

C. 制訂經濟政策時必須考慮到的因素

D. 政府對宏觀經濟的調控能力

【答案】**B**。根據否定假設法，「如果」後的內容是作者反對的，故作者贊成的是政府不應輕易改變原有的政策，即政府應持續貫徹其制訂的政策，正確答案為B。

9. 條件暗示法

【定義】當出現類似「只有……才……」、「除非……才……」、「除非……否則……」、「只要……就……」的句式時，一般強調的是緊隨「只有」、「除非」、「只要」後面的內容（即條件部分）。至於當出現「無論（不論、不管）……都（總、還、也）」的句式時，一般強調的是「都（總、還、也）」後面的結果。

【例題】以制度安排和政策導向方式表現出來的集體行為，不過是諸多個人意願與個人選擇的綜合表現。除非我們每個人都關心環境，並採取具體的行動，否則任何政府都不會有動力（或壓力）推行環保政策。即使政府制訂了完善的法規，但如果每個公民都不主動遵守，那麼，再好的法規也達不到應有的效果。

這段文字主要支持的一個觀點是：

A. 政府有責任提高全民的環保意識

B. 完善的環保法規是環保政策成敗的關鍵

C. 政府環保法規應該體現公民個人意願

D. 每個公民都應當提高自己的環保意識

【答案】D。文段中「除非」後的內容為作者所倡導的做法，即「每個人都關心環境，並採取具體的行動」，只有D項的表述與此一致。

【練習】

1. 在電腦時代，字庫就在我們每個人手邊，當我們在電腦上用文字處理軟件打出一篇文章，想要給它換上一種好看的字體時，我們其實就已經在與字庫打交道了，但很多人並不知道，目前中國漢字字庫僅421款字體，而與日本相比，漢字字庫則多達2,973款。

   這段文字的主旨是：

   A. 中國漢字字庫數量不及日本
   B. 中國漢字字庫的字體不及日本
   C. 電腦時代字庫是文字處理的必需品
   D. 中國漢字字庫的字體不夠多

2. 在法國人到達北美之前，這片土地早已住著土著居民——北美印第安人。據說他們的祖先早在3萬至1萬年前，便陸續通過亞洲東北角當時與北美仍接壤的地帶（即今日的白令海峽），來到阿拉斯加和育空地區的非凍土地帶居住下來，有的繼續向北美東南部擴展，直到大西洋沿岸和南美尖端。

   一般認為，這裡最早的土著人屬於亞洲的蒙古種族，當時亞洲還處在石器時期，這從當代加拿大發掘出來的石製器具和武器可以得到佐證。

   上文第二段的「這裡」具體指的是：

   A. 大西洋沿岸
   B. 阿拉斯加

C. 加拿大

D. 美洲

3. 德國兩家公司生產一種水底用的電腦。這種電腦使潛水員有可能在水中就能把獲得的數據輸入電腦，不必浮出水面。「水底電腦」可用於檢查海上石油平台、輸油管道、船隻，也可以用來操控機械人。

「水底電腦」的優點是：

A. 可以用來操控機械人

B. 可以用來檢查船體

C. 在水下直接把數據輸入電腦

D. 浮出水面把數據輸入電腦

4. 飲食偏好和進餐模式在人的兒童階段已形成，而兒童時代的食品選擇和飲食習慣將會對人的一生造成重要影響。收集的證據顯示，在2至11歲兒童對食物和飼料的偏好和購買「要求」，對於他們的消費習慣同樣產生不小的影響。目前的食品和飼料的構成和電視上狂轟濫炸般的兒童食品廣告，給兒童的長期身體健康帶來很大危險。

通過這段文字可看出作者的主要觀點是：

A. 兒童要養成良好的飲食習慣

B. 兒童時代的食品選擇和飲食習慣會影響一生

C. 電視廣告會影響兒童的偏好和購買「要求」

D. 垃圾食品電視廣告會影響兒童健康

5. 當年幼的藏犬長出牙齒並能撕咬時，主人就把它們放到一個沒有食物和水的封閉環境裡，讓這些幼犬自相撕咬，最後剩下一頭活著的犬，這隻犬稱為「獒」。據說十隻犬才能產生一頭獒。這種現象被稱為「犬獒效應」。

根據上述定義，下列哪句最能體現「犬獒效應」？

A. 江山代有才人出，各領風騷數百年

B. 物競天擇，適者生存

C. 不畏浮雲遮望眼，只緣身在最高層

D. 鷸蚌相爭，漁人得利

答案：

1. **D**。據提問知此題為表面主旨題。由原文「中國僅……」、「日本……則多達……」等關鍵詞可知作者對中國漢字字庫的字體數量是不滿意的。A項「數量不及日本」只是客觀事實，不能算是本文段的主旨，而表明作者態度的D項做主旨最恰當；B項偷換概念，不是字體不及日本，而是字體數量；C項在文中並未提及。故正確答案為D。

2. **D**。文段的開頭提到「北美」住著土著，接著提到有的土著來到阿拉斯加和育空地區的非凍土地帶居住下來，有的繼續向北美東南部擴展，直到大西洋沿岸和南美尖端，說明土著在整個美洲都有國居住的記錄，因此「這裡」指的是美洲。至於餘下的三個選項都過於片面，沒有涵蓋所有地區，因此排除A至C。故正確答案為D。

3. **C**。從「這種電腦使潛水員有可能在水中就能把獲得的數據輸入電腦，不必浮出水面」知道，這種「水底電腦」的優點所在。

4. **D**。本段論述的重心落在最後一句，故選D。

5. **B**。犬獒效應表達的是競爭的殘酷，在四個選項中，只有B項與之相符。A項的含義是：前人終究會被後人超越；C項的含義是：只要認識達到了一定的高度，就能透過現象看到本質，就不會被事物的假像迷惑；D項比喻雙方相持不下，而使第三者從中得利。故答案選B。

E(1). 語句排序題

【定義】

語句排序是指題目給出4至6句短句，讓考生們按照一定的思路將它排序，使之成為一個完整的段落。

【技巧】

一些考生在解答「句子排序題」時，往往會直接從題幹入手進行排列，但排完之後卻發現與選項都不一致，再次重新排列還是悲劇了。這種解法無疑是耗時耗力，結果也不盡如人意。那麼到底我們該如何解題？

第一步：觀察選項，確定首句。

1. 含指示代詞和人稱代詞的句子不作首句，例如：他/ 他們/ 這/ 這些/ 那/ 那些。

2. 成對出現的關聯詞，後者不作首句，例如：甚至/ 但是/ 而且/ 更/ 所以。

3. 含有總結性的詞語不作首句，例如：總而言之/ 綜上所述/ 由此可見/ 意味著。

4. 具體分析、無主語的句子不作首句，例如：分析原因／影響／意義的句子。

5. 表示舉例子的句子不作首句，因為前面必須先闡述觀點。

6. 生僻詞或者創新詞不作首句，因為這些詞語必須先下定義，才能出現。

第二步：找出標誌詞句，合併關聯語句。

1. 關聯詞：表示因果、轉折、遞進、條件、假設、並列、解說、順承、選擇的關聯詞，都要保持原有的詞語順序。

2. 指代詞，例如：他／他們／這／這些／那／那些，前面必須剛剛提到這個指代詞的具體指代內容。

3. 重複詞。重複出現的詞，一般應該在一起，即使不相鄰，也一定存在邏輯關係。

4. 表示時間、空間、方位詞。我們就應該把文段按照時間順序、空間順序進行排列。

5. 主體一致句，描述主體一致的句子應該相鄰。

6. 表示舉例的句子前面，一定存在這個例子所證明的內容。

【練習】

1. 選出下列句子的正確排序：

(1) 據此，海島分為大陸島、火山島、珊瑚島、衝擊島四大類型

(2) 它們萬變不離其宗，或是從大陸分離出來，或是由海底火山爆發和珊瑚蟲構造而來

(3) 盡管海島面貌千姿百態，人們仍然能找到其中的規律性

(4) 前者姓「陸」，地質構造與附近大陸相似

(5) 後者姓「海」，地質構造與大陸沒有直接關係

    A. 3-2-4-5-1

    B. 2-3-4-5-1

    C. 1-4-5-2-3

    D. 2-3-1-4-5

2. 選出下列句子的正確排序：

(1) 單純羅列史料，構不成歷史

(2) 只有在史料引導下發揮想像力，才能把歷史人物和時間的豐富內涵表現出來

(3) 歷史研究不僅需要發掘史料，而且需要史學家通過史料發揮合理想像

(4) 所謂合理想像，就是要盡可能避免不實之虛構

(5) 這是一種悖論，又難以杜絕

(6) 但是，只要想像就難以避免不實虛構出現

    A. 3-1-2-4-6-5

    B. 4-5-3-6-2-1

    C. 1-3-4-6-5-2

    D. 5-3-2-1-4-6

3. 選出下列句子的正確排序：

(1) 在丹麥、瑞士等北歐國家發現和出土的大量石斧、石製矛頭、箭頭和其他石製工具以及用樹幹造於的獨木舟便是遺證

(2) 陸地上的積冰融化後，很快就出現了苔蘚。地衣和細草，這些凍土原始植物引來了馴鹿等動物

(3) 又常年受著從西面和西南面刮來的大西洋暖濕氣流的影響，很適合生物的生長

(4) 動物又吸引居住在中歐的獵人在夏天來到北歐狩獵

(5) 北歐雖説處於高緯度地區，但這一帶正是北大西洋暖流流經的地方

(6) 這大約發生在公元前8,000年到公元前6,000年的中石器時代

    A. 6-5-3-2-4-1

    B. 5-2-3-4-1-6

    C. 5-3-2-4-6-1

    D. 6-2-4-1-5-3

答案:

1. **A**。縱觀選項，分別把（3）、（2）和（1）當做首句，而句（1）以「據此」開頭，句（2）以「它們」開頭，這兩句都是以指代詞開頭的句子，若放在首句則指代不明，所以（1）和（2）都不能作為首句，排除B、C和D，故正確選項為A。

2. **A**。其中句（5）中開頭是「這」，不能夠成為首句，故排除D項；（3）、（4）兩句當中均提到「合理想像」，句（4）是對句（3)所提及「合理想像」的解釋，句（3）在句（4）前，排除B項；觀察A、C兩個選項可以確定一下（4）、（6）、（5）三句和句（2）的順序，句（2）提出一種對策，要發揮想像力，（4）、（6）、（5）三句說的是發揮想像力出現的問題，正常邏輯應該是先提出對策，再分析次對策存在的一些問題，所以句（2）應該在（4）、（6）、（5）三句前。正確答案為A項。

3. **C**。首先通過選項可以確定句（5）或（6）一定是文段的首句。而句（6）在開頭部分出現代詞「這」，代詞出現是為了指代前文出現的內容，所以一般除了文學作品以外，代詞很少直接出現在段首，故（5）是段首句，排除A和D選項。而B、C選項的第二句和第三句分別是句（2）和（3），閱讀後可以確定句（3）與（5）的聯繫更為緊密，符合事物發展的順序，邏輯關係正確。正確答案為C。

E(2). 成語運用

【定義】

很多考生由於對一些比較生僻的成語意思不太理解，或者理解有誤，總是造成失分的情況，所以大家在平時一定要注意對成語以及成語含義的積累與運用。

【技巧】

對成語含義的理解，有些可以直接從字面得出，如「好吃懶做」、「人傑地靈」；但大多數的成語僅從字面去解釋都是不夠的，必須關注其背後的引申義、比喻義，如「胸有成竹」，是指畫竹時心裡有一幅竹子的形象，用來比喻做事前已有通盤的考慮。

要理解成語含義，第一步要解決的就是詞彙量的問題。這一點可以通過平時的積累和考前的背誦常考成語進行有效突破。

另外，理解成語過程中常出現的幾個犯錯地方，也是考生應該掌握的知識點，概括起來，主要有以下四個：望文生義、斷詞取義、忽視多義和形近音近混淆。

1. 望文生義

望文生義是指不了解某一成語的確切含義，只牽強附會從字面上理解，作出錯誤的或片面的解釋的現象。

例題：我本就對那裡的情況不熟悉，你卻硬要派我去，這不是差強人意嗎？

點撥：「差強人意」原指還能振奮人的意志，現在表示大體上能讓人滿意。這個成語在報刊中的失誤率較高。根據句意可知，此處將「差強人意」理解為「不能使人滿意」，與其本義幾乎相反，屬望文生義。

表1：常見望文生義成語

成語	意思
文不加點	「點」是刪除，不能理解為「標點」。
不刊之論	「刊」是修改，不能理解為「刊登」。
三人成虎	比喻流言惑眾、蠱惑人心，不能理解為「人多力量大」。
危言危行	「危」是正直，不能理解為「危險」。
不足為訓	「訓」是準則、典範，不能理解為「教訓」。
不名一丈	形容極其貧窮，不能理解為「不值錢」。
久假不歸	「假」是借，不能理解為「請假」。
吊民伐罪	慰問受苦的人，討伐有罪的統治者，不能理解為「治老百姓的罪」。

細大不捐	「捐」是捨棄，不能理解為「捐贈」。
罪不容誅	罪大惡極，殺了也抵不了所犯的罪惡，不能理解為「罪行還沒有達到被殺的程度」。
尾大不掉	「掉」是擺動，不能理解為「丟失」。
五風十雨	形容風調雨順，不能理解為「風雨多而成災」。
山高水低	比喻意外的不幸，多指人的死亡，不能理解為「風景很好」的意思。
曾幾何時	時間過了沒多久，不能理解為「曾經或不知何時」。
師心自用	形容自以為是、固執己見，不能理解為「善於學習借鑒，為我所用」。

2. 斷詞取義

斷詞取義是指在理解成語過程中，只斷取成語中個別語素的意義，而拋開成語其他語素的意義，從而作出錯誤或片面的解釋之現象。

例題：發展生產力是當前首當其衝的大事，是一切工作的重中之重。

點撥：「首當其衝」比喻最先受到攻擊或遭到災難。句中只斷取了「首」所含的最先、最早的意思，而拋開了「受到攻擊」或「遭受災難」之意，屬斷詞取義。

表2：常見斷詞取義成語

成語	意思
小試牛刀	不能只斷取「試」的意思，拋開「牛刀」所包含的有大才能的意思。
當仁不讓	不能只斷取「不讓」的意思，拋開「當仁」表達的面對應當做的好事的情況。
人滿為患	不能只斷取「人滿」的意思，拋開「患」表達的造成禍患的意思。
義無反顧	不能只斷取「無反顧」的意思，拋開「義」所表達的應做的事的意思。

3. 忽視多義

「多義成語」是指具有兩個或兩個以上意義的成語。對於這種成語，常出現的錯誤就是只關注了成語的常用意思，而忽視了成語的其他含義。

例題：關於金字塔和獅身人面像的種種天真、想入非非的神話和傳說，說明古埃及人有著極為豐富的想像力。

點撥：考生若只將該句中的「想入非非」理解為「胡思亂想」，則很容易據此認為其用在此處不恰當。其實，「想入非非」也可理解為「人的思想進入虛幻境界，完全脫離實際」之義。而本句中恰恰使用的是後者，是恰當的。

表3：常見多義成語

成語	意思
按圖索驥	(1)比喻按照線索尋找；(2)形容辦事機械、死板
不絕如縷	(1)比喻形勢十分危急；(2)形容聲音微弱而悠長
高山流水	(1)比喻知音難遇；(2)形容樂曲高妙
顧影自憐	(1)比喻孤獨失意的情狀；(2)形容自我欣賞
例行公事	(1)比喻按照慣例處理的公事；(2)形容只講形式，不講實效的工作
綿裡藏針	(1)比喻表面柔和、內裡剛硬；(2)形容外貌和善，但內心刻毒
瞻前顧後	(1)比喻做事謹慎；(2)形容顧慮太多、處理事情猶豫不決
平易近人	(1)比喻態度謙遜和藹，使人容易親近；(2)形容文字淺顯，通俗易懂
左右逢源	(1)比喻做事很順利；(2)形容辦事圓滑
難分難解	(1)比喻鬥爭、比賽等相持不下，難以分出勝負；(2)形容關係密切，難以分離
指手劃腳	(1)比喻說話時做出各種動作；(2)形容放肆或得意忘形的神態

4. 形近、音近混淆

有的成語與其他成語由於讀音、字形相近，在理解時極易混淆。對於這種題目，一個重要的技巧就是從成語的相異語素進行辨析，同時還要注意對相關成語進行歸納總結。

例題：球隊不孚眾望，在全場觀眾的加油助威聲下，最終以3比0戰

勝對手,在主場獲首勝。

點撥:「不孚眾望」意為不能使群眾信服。這與句中表達的取得勝利的意思不相符。此處應用「不負眾望」,表示不辜負群眾期望的意思。這兩個成語的區分可從「孚」與「負」的意思入手:「孚」意為令人信服;「負」意為辜負。

表4:常見形近、音近成語

意思相近的成語	意思
一蹴而就	比喻事情很容易做
一揮而就	形容詩文、書畫很快就寫好畫好了
不以為然	不以為是對的
不以為意	表示輕視,不放在心上
駭人聽聞	使人聽了非常震驚,所指內容通常是真實可信
聳人聽聞	故意誇大或捏造事實,使人聽了感到震驚
如虎添翼	形容強者又增添某種優勢,變得更強
為虎添翼	比喻幫助壞人,增加惡人的勢力
不容置疑	不允許加以懷疑,指絕對真實可信
不容置喙	指不允許別人插咀說話
臨危受命	在危難之時接受使命
臨危授命	遇到危難,勇於獻出自己的生命
不厭其煩	不嫌煩瑣與麻煩
不勝其煩	麻煩、囉嗦而使人受不了

【練習】

1. 中國的節日歷來與文學表現密切相關，歷代文人墨客讚美元宵節的詩句＿＿＿＿＿＿，如今讀來仍趣味無窮。

   填入劃橫線部分最恰當的是：

   A. 數不勝數
   B. 汗牛充棟
   C. 罄竹難書
   D. 不一而足

2. 同為收藏，目的和動機卻＿＿＿＿＿＿，大多數人以短期的贏利為目的，將收藏視為一本萬利的投資；另一種人源於個人愛好，作為精神的收藏。

   填入劃橫線部分最恰當的是：

   A. 天壤之別
   B. 涇渭分明
   C. 截然不同
   D. 南轅北轍

3. 明代工藝品的名字大都先強調年號，然後再強調東西本身。但景泰藍不是在景泰年間出現，而是在元代就出現了。到了景泰年間，皇帝的重視使它＿＿＿＿＿＿，因此有了今天這樣一個通俗易懂且帶有文學色彩的名字——景泰藍。

填入劃橫線部分最恰當的是：

A. 名聲大噪
B. 享譽中外
C. 聲名鵲起
D. 如日中天

4. 弗雷德·史密斯對速遞服務市場精辟獨到的分析以及他的努力、他的自信、他的非凡的領導能力，他的不可多得的膽識，特別是他_____地把全部家產投到聯邦快遞公司的勇氣和冒險精神，征服了無數精明而狡猾的風險投資大師，征服了他們口袋裡的9600萬美元。

填入劃橫線部分最恰當的是：

A. 義無反顧
B. 孤注一擲
C. 破釜沉舟
D. 鋌而走險

答案:

1. **A**。選項中四個成語都有「多」的意思,但適用對象不同。「汗牛充棟」意為藏書很多;「罄竹難書」意為罪惡很多,難以說完。這兩個成語都不能形容「讚美元宵節的詩句」多,排除B、C。「不一而足」指同類的事物或現像很多,反覆出現,不能一一列舉。詞義與句意相符,但「不一而足」一般用於句尾,故不如「數不勝數」貼切。

2. **C**。由「大多數人」和「另一種人」的對比可看出,同為收藏,目的和動機卻是差別很大的。「涇渭分明」意為界限清楚或是非分明。「南轅北轍」指行動和目的正好相反。這兩個成語均與句意不符,排除B、D。「天壤之別」比喻相隔很遠,差別很大。詞義與句意吻合,但其為名詞性成語,不能作謂語,其前應用「有」,排除A。

3. **C**。本題可從各個選項的側重點進行考慮。「名聲大噪」側重的是名聲大;「享譽中外」側重的是名聲傳得廣;「聲名鵲起」側重的是聞名的速度非常快;「如日中天」側重的是名聲正旺的狀態。本題句意是景泰藍在元代就出現了,到景泰年間因皇帝的重視才使得它迅速聞名,強調的是聞名的速度之快,故選「聲名鵲起」最合適。

4. **C**。由「勇氣和冒險精神」可看出,作者對史密斯把全部家產投到聯邦快遞公司的行為是持贊揚態度的,據此排除表貶義的B、D兩項。「義無反顧」指為了正義而勇往直前,毫不猶豫。「破釜沉舟」形容不顧一切,下定決心一拼到底。與投入全部家產的舉動相對應,「破釜沉舟」更貼切。

F(1). 字詞辨析：錯別字

【定義】

1. 錯別字定義

a.　錯字：即在字的筆畫、筆形或結構上寫錯了，例如將「猴」字寫成「候」，將「曳」字的右上角多寫了一點。

b.　別字：本來應該用某個字，卻寫成了另外一個字，例如將「戊戌政變」寫成「戊戍政變」。

2. 錯別字種類

a. 形似致誤，例如：相形見拙（絀）、如火如茶（荼）

b. 音近致誤，例如：題（提）綱、國藉（籍）、重迭（疊）

c. 義近致誤，例如：鳩佔雀（鵲）巢

d. 音、形兩近誤做，例如：急燥（躁）、貪（贓）枉法

e. 音、形、義三近誤做，例如：磨（摩）擦、甜言密（蜜）語

3. 常見錯別字例子

例如（括號內為正字）：小氣（器）、不準（准）、荀（筍）工、甘敗（拜）下風、脈博（搏）、精萃（粹）、渡（度）假村、言簡意駭（賅）、氣慨（概）、一股（鼓）作氣、粗曠（獷）

【技巧】

1. 明義記字：漢字中的象形字和指事字的字形，一般來説是有一定特殊含義的，所以最好了解它們的意思。比如「休息」的「休」是「人在樹旁歇息」，右邊是「木」字。而「體育」的「體」，是「人的根本」的意思，右邊是「本」字。「休」和「體」雖然，形似，但不會混淆。

2. 推形記字：漢字中大部分的字都是形聲字。形聲字的形旁表示漢字的意義類別，因而，我們可以抓住字的形旁，聯繫它的含義就能大致判斷出該字在句中的運用正確與否，從而辨別書寫的正誤。

3. 結構記字：有些詞尤其是四字短語，結構特殊，內部相對應的字前後意義有一定的互證聯繫，或者相反相對、相互比照，意義一致。這就可以通過其中一個字來推及另一個字，從結構上記準這類詞。例如「提心弔膽」，不能誤寫作「提心掉膽」，因為「提」和「吊」是相近的意思。

4. 詩意記字：把詞營造成富有意境的詩意文字，在獲得美的享受的同時，既容易記住字形，又能夠錘鍊自己的表意功力。例如「清秀」一詞，在任何時候，「清」和「秀」的組合都必須帶三點水，因此可描繪這樣的語句：有水分，山才朗潤（山清水秀）、眉才靈秀（眉清目秀），「美」才誕生。

【練習】

1. 請選出沒有錯字的句子：

　　A. 獲知兒子考上大學，媽媽的臉上有掩不住的喜悦。
　　B. 出版社由於經濟不景氣的關係，只好進行栽員。
　　C. 她美麗的身影成了眾所囑目的焦點。
　　D. 你怎麼可以隨便糟踢食物呢？這最要不得了！

2. 請選出沒有錯別字的句子：

　　A. 延著街市向前邁開腳步
　　B. 鐘情於慢跑這麼單調的事物
　　C. 同學要得到老師的批準，方可上洗手間
　　D. 頂著澎湖的落日餘暉

3. 請選出沒有錯別字的句子：

　　A. 他抱著破斧沉舟的決心，減重十公斤
　　B. 自從畢業以後，他就消聲匿機，同學會都不出現
　　C. 陳老師剛才的一席訓話，對我來説尤如當頭捧喝，幫我認清了自己的錯誤
　　D. 老師的教晦，我永不會忘記。

答案：

1. A。正確寫法為：B. 裁員；C. 矚目；D. 糟蹋

2. D。正確寫法為：A. 沿著；B. 鍾情；C. 批准

3. D。正確寫法為：A. 破釜沈舟；B. 銷聲匿跡；C. 當頭棒喝

F(2). 字詞辨析：簡體字運用

【定義及技巧】

簡體字（又稱「簡化字」）的幾個重點：

1. 省略：省去漢字的某部分，例如：務（务）、雲（云）、醫（
医）

2. 改形：改變漢字的外形，例如：華（华）、風（风）、眾（众）

3. 代替：以筆劃少的漢字代替筆劃多的，例如：後（后）、齣（
出）、鬥（斗）

4. 其他：有些簡化字可以獨立成字，也可以成為其它字的一部份

以下分別敘述這幾類：

1. 省略：省掉漢字某部分，我們可依省略部分歸納為下表：

省略部位	例子
上邊	雲（云）、電（电）、鬆（松）、繫（系）、麼（么）、處（处）
下邊	麗（丽）、業（业）、準（准）、禦（御）、築（筑）、屍（尸）
左邊	務（务）、錶（表）、誇（夸）、捨（舍）、鹹（咸）、餘（余）
右邊	號（号）、類（类）、殼（壳）
裡面	廣（广）、廠（厂）、氣（气）
中間	寧（宁）

外面	闢（辟）、迴（回）
重複	蟲（虫）
其他	豐（丰）、鹵（卤）、齒（齿）、飛（飞）

2. 改形：改變漢字的外形，又可細分為下表：

改變為	例子
形聲字	改換形旁：骯（肮）、願（愿）、颳（刮） 改換聲旁：膠（胶）、懼（惧）、擁（拥）、遞（递）、 癥（症） 改換形旁和聲旁：護（护）、顳（吁）、響（响）、 驚（惊）
會意字	簾（帘）、體（体）、淚（泪）、眾（众）
輪廓化	齊（齐）、農（农）、變（变）

3. 代替：用筆畫少的字取代同音或近音的字，而且還能保有原來的字意。

取代字數	例子
一個字	闆（板）、錶（表）、蔔（卜）、衝（冲）、醜（丑）、 齣（出）、疊（迭）、鬥（斗）、範（范）、穀（谷）、 颳（刮）、闔（合）、後（后）、鬍（胡）、劃（划）、 夥（伙）、幾（几）、傢（家）、薑（姜）、藉（借）、 剋（克）、睏（困）、瞭（了）、麵（面）、闢（辟）、 韆（千）、鞦（秋）、捨（舍）、瀋（沈）、鬆（松）、 塗（涂）、鹹（咸）、嚮（向）、像（象）、餘（余）、 禦（御）、鬱（郁）、願（愿）、緻（致）、製（制）、 種（种）、築（筑）、準（准）
二個字	係繫（系）、隻衹（只）
三個字	臺檯颱（台）、矇濛懞（蒙）

4. 其他：有些簡化字可以獨立成字，也可以成為其他字
的一部份

種類	例子
可以獨立成字，也可成為其他字的一部份	備（备）、貝（贝）、筆（笔）、邊（边）、參（参）、長（长）、車（车）、齒（齿）、蟲（虫）、從（从）、達（达）、黨（党）、東（东）、動（动）、爾（尔）、豐（丰）、風（风）、岡（冈）、廣（广）、龜（龟）、國（国）、華（华）、畫（画）、會（会）、幾（几）、見（见）、薦（荐）、殼（壳）、離（离）、麗（丽）、兩（两）、靈（灵）、龍（龙）、馬（马）、麥（麦）、門（门）、黽（黾）、鳥（鸟）、寧（宁）、農（农）、齊（齐）、氣（气）、遷（迁）、聖（圣）、壽（寿）、屬（属）、雙（双）、條（条）、萬（万）、韋（韦）、獻（献）、亞（亚）、業（业）、頁（页）、義（义）、藝（艺）、魚（鱼）、與（与）、雲（云）、發（发）、鹵（卤）

常見容易混淆的繁簡互換字

繁體字	簡化字	備註
纔	才	人才、才子的「才」，不可作「纔」。
衝	冲	冲洗的「冲」，不可作「衝」。
醜	丑	地支之二的「丑」，不可寫成「醜」。
範	范	姓范的「范」，不可作「範」。
後	后	皇后的「后」，不可作「後」。
鬍	胡	胡人、胡鬧的「胡」，不可作「鬍」。
迴	回	章回、回家，不可作「迴」。

【練習】

1. 請選出下面簡化字錯誤對應繁體字的選項：

   A. 築（筑）
   B. 屍（尸）
   C. 韋（韦）
   D. 纇（頁）

2. 請選出下面簡化字錯誤對應繁體字的選項：

   A. 睪（幸）
   B. 龜（龟）
   C. 穀（谷）
   D. 畫（画）

3. 請選出下面簡化字錯誤對應繁體字的選項：

   A. 後（后）
   B. 眾（从）
   C. 幹（干）
   D. 鬆（松）

答案：

1. **D**。正確對換：纇（类）。

2. **A**。正確對換：「睪」字沒有簡化字

3. **B**。正確對換：眾（众）

G(1). 句子辨析：語病句

【定義】

「語病句」是指句子在邏輯上有矛盾。通常的情況為：語序不當、搭配不當、成分殘缺或贅餘、結構混亂、表意不明或不合邏輯。

【技巧】

1. 看關聯詞語

定義：語病題多以複句形式出現，關聯詞語應該是最先關注的。關聯詞語上易出現的錯處主要是搭配不當、位置不當、不合邏輯、濫用詞語。

例子1：無論官員和人民，毫無例外，都必須遵守法律。

錯處：「無論」後只能帶由「還是」、「或」組成的詞語，而不能帶並列短語。

答案：「官員和人民」應改為「官員還是人民」

例子2：美國政府如果對進口鋼鐵實施緊急限制措施，那麼幾乎所有國家的鋼鐵業都會成為打擊對象。

錯處：當兩個分句主語不同時，第一個分句的關聯詞語應放在主語前。相反，當兩個分句的主語相同時，第一個分句的關聯詞語應放在主語後。

答案：由於例句中兩個分句主語不同，「如果」一詞的位置不當，故應放在「美國政府」前。

例子3：專家說，親子鑑定不僅「鑒」出了社會世相，也「鑒」出了血肉親情。

錯處：「不僅……也……」是表示遞進的關聯詞語，該句兩個分句內容不合邏輯。

答案：應先說「血肉親情」，後說「社會世相」。

例子4：槐茂醬菜口味獨特，深受百姓歡迎，距今已有300多年的歷史，所以仍然暢銷不衰。

錯處：前後兩個分句並無因果關係

答案：刪去「所以」一詞

2. 看句子主幹

定義：緊縮句子主幹是解題的第二步。句子主幹上易出現的「病點」主要是搭配不當、成分殘缺、句式雜糅。

例子5：用來釀製紅酒的葡萄皮中含有的白藜蘆醇，能夠提高對人體有益的高密度脂蛋白膽固醇的含量。

錯處：緊縮後的主幹是「白藜蘆醇提高含量」，謂語和賓語不搭配。

答案：「提高」一詞應改為「增加」

例子6：以網絡技術為重要支撐的知識經濟革命，極大地改變了人們的生產生活方式，加速了社會文明。

錯處：緊縮後的主幹是「知識經濟革命加速社會文明」，缺少賓語。

答案：應在「社會文明」後補上「建設」。

例子7：這屆運動會會徽、吉祥物設計的應徵者，大多是以青年師生為主。

錯處：這是句式雜糅的常見例子

答案：緊縮後的主幹是「應徵者是師生」，或「應徵者以青年師生為主」。

3. 看修飾成分

定義：解答語病題的第三步是看修飾成分。修飾成分方面易出現的「病點」是多層定語或狀語的語序不當、搭配不當、重複多餘。

例子8：鑒於她的優異成績，畢業後，水妹子留校成了該校最年輕的生物系教師。

錯處：「多層定語」的一般次序是：a. 表示領屬性的或時間、處所的；b. 指稱或數量詞；c. 動詞或動詞性短語；d. 形容詞或形容詞性短語；e. 名詞或名詞性短語。

另外，帶「的」的定語應放在不帶「的」的定語之前。

答案：「生物系」應該放在「最年輕」前。

順帶一提：多層定語的次序。（記句子──誰的──多少──怎樣的──屬性）

舉例：「她是國家隊的一位有20多年教學經驗的優秀的籃球女教練。」

語段	屬性
她是國家隊的	表示領屬性的（誰的）
一位	表示數量的（多少）
有20多年教學經驗的	詞性短語（怎樣的）

優秀的	形容詞性短語
籃球	名詞
女教練	名詞

例子9：我們順利地按照張先生頭畫的那張簡圖找到了住在美國的案件目擊者。

錯處：「多層狀語」的一般次序是：a. 表示目的或原因的（介賓短語）；b. 表示時間或處所的（名詞）；c. 表示語氣（副詞）或對象的（介賓短語）；d. 表示情態或程度的（副詞）。另外，表示對象的介賓短語一般緊挨在中心詞之前；表示總括的，一般放在「數詞＋動量詞」的後面。

答案：將「順利地」移到「簡圖」之後

順帶一提：多層狀語的次序是「時間—地點—情態—對象」

舉例：許多老師昨天在休息室裡都熱情地跟小明交談。

語段	屬性
許多老師昨天	時間
在休息室裡	地點
都熱情地	情態
跟小明交談	對象

例子10：由於採取了科技興農的策略，中國改變了長期以來糧食生產不能自給的局面。

錯處：「生產」和「自給」不搭配

答案：去掉「生產」

例子11：這句話的後面，包含了多麼豐富的無聲的潛台詞。

錯處：定語「無聲」與中心詞「潛台詞」重複

答案：刪去「無聲的」

4. 看數量短語

定義：在數量短語上易出現的病點主要是產生歧義、位置不當、倍數用錯、表約數的詞語重複。

例子12：考古科學工作者對兩千多年前在長沙馬王堆一號墓新出土的文物進行了多方面的研究。

錯處：「兩千多年前」擺放的位置不當

答案：「兩千多年前」應放在「出土的」後

例子13：和大熊貓一樣享有「國寶」之稱的丹頂鶴近年來成倍減少，目前僅存千餘隻。

錯處：數量減少不能用倍數。使用「降低」、「減少」、「縮小」等詞語時，不能用倍數。

答案：可以用分數或百分數

例子14：昨晚觀測到迸發猛烈的獅子座流星雨，目測最大強度估計超過每小時1萬顆以上。

錯處：後不能有重複表示約數的詞語，「超過」和「以上」重複

答案：應從「超過」和「以上」兩者之中去掉其中之一

順帶一提：當句中有「至少」、「最多」、「最高」、「最低」、「近」、「約」、「超過」這一類詞語時，要注意它們後面搭配的應是確數，而不能是概數。

例子15：今年以來，全廠工人幹勁十足，生產熱情高漲，產量提高到百分之二十。

錯處：數量詞在計量表述時，如果出現「增加了」、「減少了」和「增加到」、「減少到」時應區別清楚：「增加了」、「減少了」

後應接淨數，「增加到」、「減少到」後接的數量應包括底數。

答案：「提高到」應改為「提高了」

5. 看否定詞

定義：在否定詞方面易出現的「病點」，主要是多重否定造成表意相反、否定詞和反問句連用造成表意相反、否定詞和帶有否定意義的動詞連用，造成表意相反、位置不當。

例子16：其實，只要部分觀眾適應了字幕版的放映方式，根本就沒有必要不因為配音這一環節而造成不必要的資金消耗。

錯處：多重否定不當，應去掉其中一重否定。

答案：將「不因為」的「不」字去掉

例子17：睡眠三忌：一忌睡前不可惱怒，二忌睡前不可飽食，三忌臥處不可當風。

錯處：「忌」含有否定意義，再出現否定詞，造成否定混亂。

答案：應去掉三個「不可」

順帶一提：含有否定意味的一類詞常見的有「防止」、「禁止」、

「反對」、「切忌」、「拒絕 」、「杜絕」、「避免」、「阻擋」
等，在病句審查時要特別注意。另外「否則」後面不能接「如果（
若）不這樣」的句子，不然就犯了重複的毛病。

6. 看兩面詞

定義：兩面詞指的是句子中出現的諸如「能否」、「是否」、「成
敗」和「好壞」之類的詞語。在兩面詞上易出現的病點主要是兩面
對一面，或一面對兩面的不照應。

**例子18：我們能不能培養出「四有」新人，是關係到國家前途命運
的大事，也是教育戰線的根本任務。**

錯處：「能不能」表兩面，後面陳述的只表一面，不能照應，該句
犯了兩面對一面的錯誤。

**例子19：照片拍得好，詩歌寫得有味，是由一個人的思想認識、藝
術修養的高低所決定的。**

錯處：「好」、「有味」和「高低」不照應，該句犯了一面對兩面
的錯誤。

7. 看介詞

定義：在介詞上易出現的病點主要是缺少主語、主客顛倒、搭配不當、結構混亂、漏用濫用、造成歧義。

例子20：經過老師耐心的教育，終於使我醒悟過來，我真的錯了。

錯處：介詞開頭的句子，要特別注意是否湮沒了主語，造成了主語殘缺的毛病。順帶一提：出現在句首的介詞常見的有：「通過」、「經過」、「由於」、「對於」、「為了」等，而且句中常伴有「使」字出現。

答案：「經過」使主語殘缺，應去掉。

例子21：根據法庭對「黑哨事件」的調查結果和司法機構給出的書面材料看，他是在未被採取強制措施時交代了自己的罪行的。

錯處：當介詞成對出現時，容易出現搭配不當的毛病。所以要特別注意由介詞和它後面的賓語組成的介賓短語是否完整。這類介詞常見的有以下：「在……上（下）」、「從……中」、「從……出發」、「以……為中心」、「以……為代價」、「以……為主」、「當……時」、「由……組成」等。這類介詞常常與句式雜糅又有著千絲萬縷的聯繫，審查病句時要特別注意。

答案：「根據」不能和「看」搭配，應把「根據」改為「從」。

例子22：學校自從調整了作息時間後，許多學生由於開始不習慣，上課經常遲到。

錯處：首句結構混亂

答案：將「自從」應移到「學校」前，使首句做狀語成分。

例子23：「貧鈾」對人的危害，主要體現在肝腎的破壞上，並有可能由此導致人的死亡。

錯處：漏用介詞「對」

答案：應將「對」加在「肝腎」前

例子24：在對WTO問題的關注上，過去主要集中在行業、企業等方面所面臨的壓力和挑戰上。

錯處：首句的「在」和「上」，多餘。

答案：應去掉「在」和「上」

8. 看代詞

定義：使用代詞易出現的「病點」，主要是指代不明造成歧義。

例子25：這次採訪的外國球隊給我們的青年隊員上了很好的一課，恐怕他們終生都不會忘記這次比賽。

錯處：「他們」指代不明，可指「外國球隊」，也可指「我們的青年隊員」。

答案：將「他們」寫成「我們的青年隊員」

例子26：原書法協會主席陳先生看過李小姐的作品後，稱讚其「深得神韻，獨有所長」。

錯處：「其」指代不明，可指「李小姐」，也可指「作品」。

答案：將「其」換成「其作品」

9. 副詞

定義：一些表示時間、地點、範圍、程度或心理活動的副詞作修飾語時，要特別注意與後面的中心語是否有語意重複毛病。

例子27：專家們特別指出：全城免費WiFi是有關當局目前的當務之急。

錯處：「目前」與「當務之急」語意重複

答案：刪去「目前」

例子28：A牌子的襯衫無論在款式上、質量上，還是包裝上，都可以堪稱全國一流。

錯處：像這類短語常見的有：真知灼見的意見、常常屢見不鮮、來自於、國際間、這其中、此個中、令人堪憂、被應邀、過分溺愛、過分苛求、過高奢望、多年夙願、過分溢美之詞、隨便苟同、好像如芒在背、特別窮凶極惡、三令五申地強調、非常奇缺、迅速立竿見影、顯得相形見絀、您的垂詢、不可回測、倆個、杜絕不要、切忌不要、避免不要、非法走私活動、基本上根除、壞毛病、目的是為了、原因是因為、不透明的暗箱操作、從來沒有過的空前盛況……

答案：「可以」與「堪稱」語意重複，應刪去「可以」。

【練習】

1. 下列各句中，沒有語病的一句是：

A. 文藝之於民俗是傳承更是發展，從理論上講要想在文藝話語中找不到民俗真的很難。同樣，文藝對民俗的傳承也愈加顯得更加重要。

B. 《中國通史》共拍攝了100集，再現了中國上下5,000年的浩瀚歷史圖景和變遷，全面而系統地展示了豐富燦爛的包括敦煌文化在內的中華文明。

C. 天越來越陰沉，大暴雨馬上就要降臨了，路人都行色匆匆，可修車人倒顯得非常鎮靜。

D. 在北京這個大的城市背景下，在已成定制的傳統建築空間佈局的住宅形式內，世世代代的北京人演繹著國都的輝煌和市井的喧囂。

2. 下列各句中，沒有語病的一句是：

A. 中國印章已有兩千多年歷史，它由實用逐步發展成為一種具有獨特審美的藝術門類，受到文人、書畫家和收藏家的推崇。

B. 創新研究性大學必須建立更加開放的辦學方向，深化與世界各國的著名高校和學術組織全方位、多層次的實質性合作交流，鞏固和加強各種類型合作平台的建設。

C. 空談之風四處蔓延，甚至影響到了孩子們，作文中的「假大空」和電視鏡頭中的「標準化表情與表達」，毒化了原本樸實的社會風氣，下一代的失真與失實成為常態，讓人為之擔憂。

D. 微波具有乾燥、殺菌等多種功能，廣泛用於食品。它與收音機所用的電波在本質上是同一種東西，使用微波爐致癌目前並無準確數據支持。

3. 下列各句中，沒有語病的一句是：

A. 微生物超標的原因可能是由於從業人員不注意個人衛生，或生產、運輸、貯存、銷售環節控制不嚴導致產品受到污染。

B. 隨著國家對西部經濟社會發展的支持力度加大，加上西部地區新一輪快速增長勢頭正在方興未艾，預計2012年西部地區仍將是全國經濟增長最快的地區。

C. 近期，美國三藩市加州大學華裔博士潘登發現了導致生物體衰老的重要基因，這一研究成果震驚學界，其論文被衰老機制及老年疾病研究的最權威期刊《Aging Cell》刊登。

D.《理財周報》發佈2012年3,000個內地家族財富榜，「三一重工」梁穩根家族406億元身家首次蟬聯內地首富，吳亞軍家族財富404億元，排名第二。

4. 下列各句中，沒有語病的一句是：

A. 進入幼師的門檻太低，其根本原因在於國家把學齡前教育作為基礎教育的「包袱」被甩掉了，最終導致各類學前教育機構成為少數人的斂財渠道。

B. 在資溪縣慘遭殺戮的不僅僅是彌猴，野生動物的日子同樣不好過。它們既要提防飛來的子彈，還要小心獵人佈下的重重陷阱。

C. 當野生植物到達基地後，「馴化師」開始通過人工模擬自然環境，改良植物土壤環境和溫室氣候條件，對這些植物開展「馴化」，讓它們適應城市生活。

D. 湖北省襄陽城，很多金庸迷都非常熟悉，那裡是一座軍事要塞，有《射鵰英雄傳》郭靖和黃蓉死守襄陽城抗元的故事。

答案:

1. **D**。觀乎A項,語意重複,「愈加」與「更加」重複,去掉「更加」;B項,語序不當,把「豐富燦爛的」移到「中華文明」之前;C項,有歧義,「修車人」既可指「修車師傅」,也可指「車主」。

2. **C**。觀乎A項,成分殘缺,應為「具有獨特審美價值的藝術門類」;B項,搭配不當,應為「建立更加開放的辦學模式」;D項,成分殘缺,應為「廣泛用於食品加工」,「使用微波爐致癌的說法目前並無準確數據支持」。

3. **C**。觀乎A項,句式雜糅,把「由於」或「的原因」去掉;B項,語意重複,把「正在」去掉;D項,介詞殘缺,在「三一重工梁穩根家族」之後加「以」。

4. **D**。觀乎A項,句式雜糅,可刪除「被」。B項,不合邏輯,彌猴也是野生動物。應在「野生動物」前加「其它的」。C項,搭配不當,應為「改善土壤環境和溫室氣候條件」。

G(2). 句子辨析：邏輯錯誤

【定義】

邏輯錯誤是指不恰當的推理。

【技巧】

1. 同語反覆

例如：樂觀主義者就是樂觀地對待生活的人。

2. 循環定義

例如：如果把丈夫定義為妻子的愛人，那麼，妻子就是丈夫的愛人。

3. 概念不當並列

例如：音樂分為古典音樂、鄉村音樂、流行音樂和民族音樂等。

4. 偷換概念

例如：司馬光說：「我要去看花燈。」

司馬光夫人說：「家中這麼多燈，何必去看？」

司馬光夫人說：「我要去看遊人。」

司馬光說：「家中這麼多人，何必出去看？」

5. 轉移論題

例如：「我以為中學生沒有必要學習地理。整個國家的地形和位置完全可以和這個國家的歷史同時學習。我主張可以把歷史課和地理課合並併，這樣對學生是方便的。」

6. 自相矛盾

例如：「這件事情我沒有問過，只是側面了解一下情況，提點意見，僅供參考。」

7. 兩皆不可

例如：「這篇文章的觀點不能說是全面的，也不能說是片面的。」

8. 以偏概全

這是不正確構造，簡單枚舉歸納推理時出現的邏輯錯誤。

9. 循環論證

這種錯誤發生在一個論證中，論據的證明需要依賴前提的情況。

10. 倒置因果

例如：為了加快中國的發展，必須大力發展航天工業。因為在發達國家，航天工業發展很快。

【練習】

選出以下沒有犯邏輯錯誤的句子：

1. 平均而言，現在受過教育的人的讀書時間明顯少於50年前受過教育的人的讀書時間，但是，現在每天的售書冊數卻比50年前增加了很多。

   以下各項陳述都有助於解釋上述現象，除了：

   A. 現在受過教育的人比50年前受過教育的人的數量大大增加
   B. 與現在相比，50年前的人更喜歡從圖書館借閱圖書
   C. 與現在相比，50年前的人更喜歡通過大量藏書來顯示其良好的教育和地位
   D. 現在的書往往比50年前的書更薄，也更容易讀

2. 以下是一段由教授與學生的對話：

   教授：在長子繼承權的原則下，男人的第一個妻子生下的第一個男性嬰兒總是首先有繼承家庭財產的權利。

   學生：那不正確。休斯敦夫人是其父惟一妻子的惟一活著的孩子，她繼承了他的所有遺產。

   學生誤解了教授的意思，他理解為：

   A. 男人可以是孩子的父親
   B. 女兒不能算第一個出生的孩子
   C. 只有兒子才能繼承財產
   D. 私生子女不能繼承財產

3. 因為太陽一直照常升起，所以明天太陽還會照常升起。

以下哪項與上述推理的邏輯錯誤不一致？

A. 巧婦難為無米之炊

B. 冬天來了，春天還會遠嗎？

C. 這麼如雷貫耳的名字，想必你已經久仰了。

D. 從來沒有人躲得過他的飛刀，你別去送死了！

答案：

1. **C**。A、B、D三項分別從現在受教育的人數量比50年前多因此消費更多圖書、50年前人們喜歡通過從圖書館借閱而不是買書的方式讀書、現在的書比50年前的書更具有易讀性因此人們可以用更少的時間讀更多的書等方面合理地解釋了題幹中的矛盾現象。C指出，50年前的人更喜歡通過大量藏書來顯示其良好的教育和品位，也就是説50年前的人更喜歡買書，這就不能解釋「現在每天的售書冊數卻比50年前增加了很多」這一現象，故選C項。

2. **C**。根據教授的結論，長子繼承權是特定男性嬰兒的權利，但並不排除女兒也有可能繼承財產，學生忽略了這個可能，所以造成了誤解。在只有女兒的情況下，女兒當然具有繼承財產的權利，這並不會對長子繼承權構成反駁，答案為C。

3. **A**。文段翻譯成三段論就是：以前的太陽都是照常升起的，明天的太陽也是太陽，所以明天的太陽也會是照常升起的。其中「以前的太陽」與「太陽」不是一個概念，犯了「四概念」的錯誤。A項暗含的前提是「無米不成炊」，翻譯成三段論是：因為沒米做不成飯，現在沒米，所以做不成飯。這是大前提錯誤，與文段不一致。B項假設的前提是「以前的冬天之後一定是春天」，翻譯成三段論是：因為以前的冬天之後是春天，這個冬天也是冬天，所以這個冬天之後也是春天。C項「如雷貫耳」比喻名聲人，也就是大部分人都知道，這句話翻譯成三段論是：因為大部分人都知道這個名字，你也是人，所以你也知道。D項翻譯成三段論是：因為以前的人都躲不過他的飛刀，你也是人，所以你也躲不過。B、C、D都是犯了「四概念」的錯誤，所以本題選A。

H. Verbal Reasoning (English)

【定義】

Verbal Reasoning（語文推理）是根據一篇短文，來判斷題幹信息正確與否，主要考察考生的英語閱讀能力和邏輯判斷能力。此類測試的答案分三種：True（正確）、False（錯誤）和Can't tell（未提及）。

【技巧】

快速判斷True、False和Can't tell的方法：

a. True的特點

(1) 題目是原文的同義表達，通常用同義詞或同義結構。例如：

【原文】Few are more than five years old.

【題目】Most are less than five years old.

【答案】題目與原文是同義結構，故答案為True。

(2) 題目是根據原文中的幾句話做出推斷或歸納。不推斷不行，但有時部分考生會走入另一個極端，即自行推理或過度推理。例如：

【原文】Compare our admission inclusive fare and see how much you save. Cheapest is not the best and value for money is guaranteed. If you compare our bargain Daybreak fares, beware - most of our competitors do not offer an all inclusive fare.

Daybreak fares are more expensive than most of their competitors.

【答案】雖然文章沒有直接提到的費用比絕大多數的競爭對手都昂貴，但從原文幾句話中可以推斷出Daybreak和絕大多數的競爭對手相比收費更高，但服務的項目要更全。與題目的意思一致，故答案為True。

b. False的特點

(1) 題目與原文直接相反。常用於反義詞、not加同義詞及反義結構。例如：

【原文】A species becomes extinct when the last individual dies.

【題目】A species is said to be extinct when only one individual exists.

【答案】可以看出題目與原文是反義結構。原文説一個物種死光，才叫滅絕，而題目説還有一個個體存活，就叫滅絕。由於題目與原文直接相反，故答案為False。

(2) 原文是多個條件並列，題目是其中一個條件（出現must或only）。原文是兩個或多個情形（通常是兩種情形）都可以，常有both... and、and、or、also等詞。題目是「必須」或「只有」其中一個情況，常有must及only等詞。例句：

【原文】Booking in advance is strongly recommended as all Daybreak tours are subject to demand. Subject to availability, stand by tickets can be purchased from the driver.

【題目】Tickets must be bought in advance from an authorized Daybreak agent.

【答案】原文是提前預定、直接向司機購買都可以，是多個條件的並列。題目是必須提前預定，是必須其中一個情況。答案為False。

(3) 原文為人們對與於某樣事物的理論或感覺，題目則強調是客觀事實或已被證明。原文強調是一種「理論」或「感覺」，常有、及等詞。題目強調是一種「事實」，常有fact及prove等詞。例如：

【原文】But generally winter sports were felt to be too specialized.

【題目】The Antwerp Games proved that winter sports were too specialized.

【答案】原文中有feel，強調是感覺。題目中有prove，強調是事實。答案為False。

(4)原文和題目中使用了表示不同範圍、頻率、可能性的詞。原文中常用many、sometimes、unlikely等詞。題目中常用all、usually、always和impossible等詞。例如：

【原文】Frogs are sometimes poisonous.

【題目】Frogs are usually poisonous.

【答案】原文中有sometimes，強調是「有時」。題目中有usually，強調是「通常」。答案為False。

(5) 原文中包含條件狀語，題目中去掉條件成份。原文中包含條件狀語，如if、unless或if not也可能是用介詞短語表示條件狀語如in、with、but for或exept for。題目中去掉了這些表示條件狀語的成份。這時，答案應為False。例如：

【原文】The Internet has often been criticized by the media as a hazardous tool in the hands of young computer users.

【題目】The media has often criticized the Internet because it is dangerous.

【答案】原文中有表示條件狀語的介詞短語in the hands of young computer users，題目將其去掉了。答案為 False。

c. Can't tell

(1) 題目中的某些內容在原文中沒有提及。題目中的某些內容在原文中找不到依據。

(2) 題目中涉及的範圍小於原文涉及的範圍，也就是更具體。原文涉及一個較大範圍的範疇，而題目是一個具體概念。也就是說，題目中涉及的範圍比原文要小。例如：

【原文】Our computer club provides printer.

【題目】Our computer club provides color printer.

【答案】題目中涉及的概念「比原文中涉及的概念」要小。換句話說，計算機俱樂部提供打印機，但是是彩色還是黑白的，不知道或有可能，文章中沒有給出進一步的信息。答案為Not Given。

(3) 原文是某人的目標、目的、想法、願望、保證、發誓等，題目是事實。原文中常用aim、purpose、promise、swear、vow等詞。題

目中用實意動詞。例如:

【原文】He vowed he would never come back.

【題目】He never came back.

【答案】原文中說他發誓將永不回來,但實際怎麼樣,不知道。也可能他違背了自己的誓言。答案為Not Given。

(4) 題目中有比較級,原文中沒有比較。例子如下:

【原文】In Sydney, a vast array of ethnic and local restaurants can be found to suit all palates and pockets.

【題目】There is now a greater variety of restaurants to choose from in Sydney than in the past.

【答案】原文中提到了悉尼有各種各樣的餐館,但並沒有與過去相比。答案為Not Given。

【練習】

Passage 1

The project was ambitious in its size, complexity, triparty nature, and in its pioneering of the Private Finance Initiative. This difficulty was unavoidable and contributed to the project's failure. However, a more thorough estimate of the unknown difficulties and timescales would have enabled the Department to better prepare for the project, and increase its chance of success.

In December 1997 XSoft indicated they needed time to complete the project, which should have been inevitable. If the Department knew from the start how long the project would take, it is questionable whether they would have considered inception, especially considering the implications of delay on the overall profitability for the venture.

1. If more care had been put into estimating the difficulties, it is less likely the project would have failed.

2. XSoft witheld information from the Department regarding how long the project would take.

3. The Department's profits were dependent upon how long the project took.

Passage 2

Ever since the gun's invention it has been changing the world in many different ways. Many of the developments in gun design have been brought about by man's desire to protect himself, and the challenge of inventing bigger and more accurate weapons.

Each time there has been a major innovation in the development of the gun, there has been a profound effect on the world. The gun helped in the exploration of the world, it has also helped in the development of society as we know it.

4. The gun was invented because the human race needs to protect themselves.

5. Guns are the reason our society is the way it is today.

6. Financial incentives had no part to play in the development of the gun.

Passage 3

Being socially responsible is acting ethically and showing integrity. It directly affects our quality of life through such issues as human rights, working conditions, the environment, and corruption. It has traditionally been the sole responsibility of governments to police unethical behavior. However, the public have realized the influence of corporations and, over the last ten years, the level of voluntary corporate social responsibility initiatives that dictate the actions of corporations has increased.

7. The ethical actions of corporations has changed over the last ten years.

8. Corporations can influence the public's quality of life.

9. Traditionally, the government have relied upon only the large corporations to help drive corporate social responsibility, whilst they concentrated on the smaller corporations.

答案：

1. **Answer: True.**

   Since more thorough can be considered equivalent to giving more care, this is stated as true in the following excerpt "a more thorough estimation of the unknown difficulties and timescales would···increase its chance of success".

2. **Answer: Can't tell.**

   We are told XSoft requested more time, and the passage implies the Department did not know how long it would take at the beginning, but the passage does not tell us if XSoft did or did not withhold time information.

3. **Answer: True**

   We are told that there were "implications of delay on overall profitability for the venture".

4. **Answer: Can't tell**

   The passage does not say how or why the gun was invented. It does say that some changes to the gun's design have been because humans want to protect themselves, but the passage does not say how or why the gun was first invented.

5. **Answer: False**

   The gun "has also helped in the development of society as we know it". The word help implies it is not the only contributor and is therefore not the reason our society is the way it is today.

6. **Answer: Can't tell**

   The passage does not mention financial incentives or economic gain, so we cannot tell from information in the passage alone.

7. **Answer: True**

   The passage says that the public have caused corporations to alter their ethical actions: "over the last ten years, the level of voluntary corporate social responsibility initiatives that dictate the actions of corporations has increased."

8. **Answer: True**

   This one is slightly less obvious. We are first told that being socially

   responsible directly affects our quality of life. Then we are told that corporate social responsibility dictates the actions of corporations. So following that logic, corporations must be able to affect our quality of life.

9. **Answer: False**

   We are told that "traditionally it has been the sole responsibility of the government to police unethical behavior", meaning traditionally no corporation played a part; not even large corporations.

I. Data Sufficiency

1. What percent of a group of people are women with red hair?

(1) Of the women in the group, 5 percent have red hair.

(2) Of the men in the group, 10 percent have red hair.

題目問：「紅頭髮女人在這群人當中佔的百分比？」

想要知道百分比，首先需要知道人群的總數，以及紅髮女人的數目。條件(1)說女人中有5%的女人為紅髮，條件(2)說有10%的男人是紅髮。兩個條件都得不出紅髮女人所佔的比例，故答案選E——兩個條件都不充分。

2. In a certain class, one student is to be selected at random to read. What is the probability that a boy will read?

(1) Two-thirds of the students in the class are boys.

(2) Ten of the students in the class are girls.

題目問：「課堂上隨機選擇一個學生閱讀，那麼選到男生閱讀的概率是多少？」

考生需要知道男生所佔全班人數的比率，才能知道男生閱讀的概率，故條件(1)充分，概率是三分之二。條件(2)只說明女生有10個，跟題幹的男生沒有任何關係，故答案選A。

3. If the two floors in a certain building are 9 feet apart, how many steps are there in a set of stairs that extends from the first floor to the second floor of the building?

(1) Each step is foot high.

(2) Each step is 1 foot wide.

題目問：「在一座建築物內，兩層樓梯間的距離9呎。那麼，從第一個樓梯到第二個樓梯，共有多少級台階？」

考生只需要知道每個台階的高度，就可以計算出台階的數目。條件(1)說明每級台階有0.75呎高，條件充分。條件(2)表示每級台階都是闊1呎，與題幹沒有相關性，故選A——只有條件(1)符合。

4. In a college, the number of students enrolled in both a chemistry course and a biology course is how much less than the number of students enrolled in neither?

(1) In College X there are 60 students enrolled in a chemistry course.

(2) In College X there are 85 students enrolled in a biology course.

題目問：「院校內既學習化學，又學習生物的學生數目，比兩個科目都不學的學生數量少幾多？」

考生需要知道兩樣都學的人數多少，兩樣都不學的人數又有多少。條件(1)指出上有60位化學堂的學生，條件(2)表示讀生物科的有85人。兩個條件不論怎麼配搭都求不出來，故答案是E──兩個條件都不符合。

5. If n is an integer, is n+1 odd?

(1) n+2 is an even integer

(2) n-1 is an odd integer

題目問：「如果n是整數，那麼n+1是奇數嗎？」

條件(1)說n+2是偶數，條件(2)說n-1是奇數。考生想要知道n+1是奇數，那麼滿足n是偶數就可以。條件(1)得出n是偶數，條件充分。條件(2)得出n是偶數條件也滿足，所以兩個條件都是單獨滿足，故答案選D。

【練習】

1. A die is rolled randomly on to a circular board with a triangle inscribed in the circle (All three vertices of the triangle are on the circumference of the circle).What is the probability that the die comes to rest outside the triangular region?

   (1) The hypotenuse of the triangle is a diameter of the circle.
   (2) The radius of the circle is 2 units, and the area of the triangle is 4 square units.

   A. Statement (1) alone is sufficient, but statement (2) alone is not sufficient to answer the question.
   B. Statement (2) alone is sufficient, but statement (1) alone is not sufficient to answer the question.
   C. Both statements taken together are sufficient to answer the question, but neither statement alone is sufficient.
   D. Each statement alone is sufficient.
   E. Statements (1) and (2) together are not sufficient, and additional data is needed to answer the question.

2. In what year was Mary born?

   (1) Mary's daughter was born in 1960 when Mary was 28 years old.

   (2) Mary's birthday and her daughter's birthday are exactly six months apart.

   A. Statement (1) alone is sufficient, but statement (2) alone is not sufficient to answer the question.

B. Statement (2) alone is sufficient, but statement (1) alone is not sufficient to answer the question.

C. Both statements taken together are sufficient to answer the question, but neither statement alone is sufficient.

D. Each statement alone is sufficient.

E. Statements (1) and (2) together are not sufficient, and additional data is needed to answer the question.

3. Is $xy < 15$?

(1) $0.5 < x < 1$, and $y^2 = 144$

(2) $x < 3$, $y < 5$

A. Statement (1) alone is sufficient, but statement (2) alone is not sufficient to answer the question.

B. Statement (2) alone is sufficient, but statement (1) alone is not sufficient to answer the question.

C. Both statements taken together are sufficient to answer the question, but neither statement alone is sufficient.

D. Each statement alone is sufficient.

E. Statements (1) and (2) together are not sufficient, and additional data is needed to answer the question.

4. x = 1.24d5. If d is the thousandth's digit in the decimal above, what is the value of x when rounded to the nearest hundredth?

(1) $x < \dfrac{5}{4}$ (2) d < 5

A. Statement (1) alone is sufficient, but statement (2) alone is not sufficient to answer the question.

B. Statement (2) alone is sufficient, but statement (1) alone is not sufficient to answer the question.

C. Both statements taken together are sufficient to answer the question, but neither statement alone is sufficient.

D. Each statement alone is sufficient.

E. Statements (1) and (2) together are not sufficient, and additional data is needed to answer the question.

5. Are the integers x, y and z consecutive?

(1) The arithmetic mean (average) of x, y and z is y.

(2) y-x = z-y

A. Statement (1) alone is sufficient, but statement (2) alone is not sufficient to answer the question.

B. Statement (2) alone is sufficient, but statement (1) alone is not sufficient to answer the question.

C. Both statements taken together are sufficient to answer the question, but neither statement alone is sufficient.

D. Each statement alone is sufficient.

E. Statements (1) and (2) together are not sufficient, and additional data is needed to answer the question.

Answer

1. **B**   The find the probability we need to know the ratio of the area inside the triangle to the area outside. Statement (1) tells us that the triangle lies in one semi-circle and is a right angled triangle, but does not allow us to fix the exact area of the triangle. Thus the answer cannot be A or D. Statement (2) allows us to find the area of the circle and the area of the triangle is given, and so is sufficient to find the ratio of the areas and the answer is B.

2. **E**   Someone who is 28 in one particular year could have become 28 that same year or have become 28 the previous year, depending on the month of birth, and so the information in statement (1) is not sufficient to answer the question of whether Mary was born in 1931 or 1932. The information in statement (2) places the birthdays six months apart, but give no information on age hence the answer must be C or E. Even combined with statement (1) the information in statement (2) leaves open the possibility of a 1931 or 1932 birth year for Mary and so the answer is E.

3. **A**   From statement (1) we can see that y must be 7 or -7 and x is a positive fraction. A fraction of 7 or of -7 will always be less than 15 and so statement (1) is sufficient and the answer must be A or D. Statement (2) alone does not answer the

questions definitely because x and y could both be positive numbers with a product less than 15 of they could both be negative with large absolute values so that their product is greater than 15. The answer is A.

4. **D** When rounded to the nearest hundredth x can be 1.24 or 1.25, depending on whether we round up or down. If d is ≥ 5 the value of x when rounded up will be 1.25. Statement (1) says that x is less than 1.25. Hence we know we must round down to 1.24 and the answer must be A or D. If d is less than 5, as stated in statement (2), we know we have to round down to 1.24 and so the answer is D.

5. **E** The mean of three numbers will equal the middle number for any set of evenly spaced numbers (1, 2, 3 or 2, 4, 6, or -1, -4, -7 for example) and so we cannot assume that x, y and z are consecutive. Hence the answer cannot be A or D. If x, y and z were 2, 4, and 6, for example, the equation in statement (2) would be valid, so once again the numbers do not have to be consecutive, and the answer cannot be B. From the numbers we have just substituted, we should be able to see that putting statements (1) and (2) together will still not give a situation in which the numbers are always consecutive or always not consecutive, and so the best answer is E.

J. Interpretation of Tables and Graphs

Table Chart

Expenditures of a Company (in Lakh Rupees) per Annum Over the given Years.

Year	Item of Expenditure				
	Salary	Fuel and Transport	Bonus	Interest on Loans	Taxes
2013	288	98	3.00	23.4	83
2014	342	112	2.52	32.5	108
2015	324	101	3.84	41.6	74
2016	336	133	3.68	36.4	88
2017	420	142	3.96	49.4	98

1. What is the average amount of interest per year which the company had to pay during this period?

    A. Rs.32.43 lakhs

    B. Rs.33.72 lakhs

    C. Rs.34.18 lakhs

    D. Rs.36.66 lakhs

Answer: D

Average amount of interest paid by the Company during the given period

$$= Rs.\left[\frac{(23.4+32.5+41.6+36.4+49.4)}{5}\right] \text{ lakhs}$$

$$= Rs.\left[\frac{183.3}{5}\right] \text{ lakhs}$$

$= Rs.36.66$ lakhs

2. The total amount of bonus paid by the company during the given period is approximately what percent of the total amount of salary paid during this period?

    A. 0.1%

    B. 0.5%

    C. 1%

    D. 1.25%

Answer: C

Required percentage = $[\frac{(3.00+2.52+3.84+3.68+3.96)}{(288+342+324+336+420)} \times 100]\%$

$= [\frac{17}{1710} \times 100]\%$

= approximately 1%

3. Total expenditure on all these items in 2013 was approximately what percent of the total expenditure in 2017?

    A. 62%

    B. 66%

    C. 69%

    D. 71%

Answer: C

Required percentage = $[\frac{(288+98+3.00+23.4+83)}{(420+142+3.96+49.4+98)} \times 100]\%$

$$= [\frac{495.4}{713.36} \times 100] \%$$

= approximately 69.45%

4. The total expenditure of the company over these items during the year 2015 is?

    A. Rs.544.44 lakhs

    B. Rs.501.11 lakhs

    C. Rs.446.46 lakhs

    D. Rs.478.87 lakhs

Answer: A

Total expenditure of the Company during 2015

= Rs.(324+101+3.84+41.6+74) lakhs

= Rs.544.44 lakhs

Bar Chart

The bar graph given below shows the sales of books (in thousand number) from six branches of a publishing company during two consecutive years 2016 and 2017.

Sales of Books (in thousand numbers) from Six Branches: B1, B2, B3, B4, B5 and B6 of a publishing Company in 2016 and 2017.

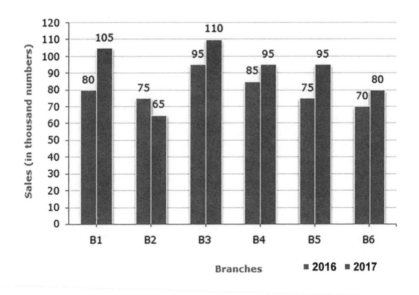

1. What is the ratio of the total sales of branch B2 for both years to the total sales of branch B4 for both years?

    A. 2:3

    B. 3:5

    C. 4:5

    D. 7:9

Answer: D

Required ratio $= \dfrac{(75+65)}{(85+95)} = \dfrac{140}{180} = 7:9$

2. Total sales of branch B6 for both the years is what percent of the total sales of branches B3 for both the years?

    A. 68.54%

    B. 71.11%

    C. 73.17%

    D. 75.55%

Answer: C

Required percentage $= [\dfrac{(70+80)}{(95+110)} \times 100]\%$

$= [\dfrac{150}{205} \times 100]\%$

$= 73.17\%$

3. What percent of the average sales of branches B1, B2 and B3 in 2017 is the average sales of branches B1, B3 and B6 in 2016?

   A. 75%

   B. 77.5%

   C. 82.5%

   D. 87.5%

Answer: D

Average sales (in thousand number) of branches B1, B3 and B6 in 2016

$$= \frac{1}{3} \times (80+95+70) = (\frac{245}{3})$$

Average sales (in thousand number) of branches B1, B2 and B3 in 2001

$$= \frac{1}{3} \times (105+65+110) = (\frac{280}{3})$$

Therefore, required percentage $= [\frac{(245/3)}{(280/3)} \times 100]\%$

$= (\frac{245}{280} \times 100)\% = 87.5\%$

Line Chart

Study the following line graph and answer the questions.

Exports from three companies over the years (in Rs.crore)

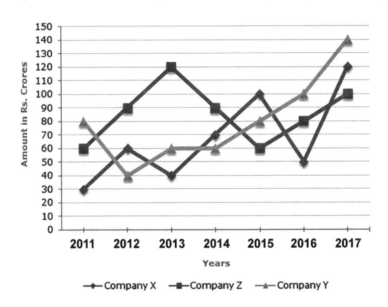

1. For which of the following pairs of years the total exports from the three Companies together are equal?

   A. 2013 and 2016

   B. 2014 and 2016

   C. 2015 and 2016

   D. 2013 and 2014

Answer: D

Total exports of the three companies X, Y and Z together, during various years are:

In 2011, Rs.(30+80+60) crores = Rs.170 crores

In 2012, Rs.(60+40+90) crores = Rs.190 crores

In 2013, Rs.(40+60+120) crores = Rs.220 crores

In 2014, Rs.(70+60+90) crores = Rs.220 crores

In 2015, Rs.(100+80+60) crores = Rs.240 crores

In 2016, Rs.(50+100+80) crores = Rs.230 crores

In 2017, Rs.(120+140+100) crores = Rs.360 crores

Clearly, the total exports of the three companies X, Y and Z together are same during the years 2013 and 2014.

2. Average annual exports during the given period for company Y is approximately what percent of the average annual exports for company Z?

   A. 87.12%
   B. 89.64%
   C. 91.21%
   D. 93.33%

Answer: D

Analysis of the graph: From the graph it is clear that,

1. The amount of exports of company X (in crore Rs.) in the years 2011-2017 are 30, 60, 40, 70, 100, 50 and 120 respectively.

2. The amount of exports of company Y (in crore Rs.) in the years 2011-2017 are 80, 40, 60, 60, 80, 100 and 140 respectively.

3. The amount of exports of company Z (in crore Rs.) in the years 2011-2017 are 60, 90, 120, 90, 60, 80 and 100 respectively.

Average annual exports (in Rs.crore) of Company Y during the given period

$$= \frac{1}{7} \times (80+40+60+60+80+100+140) = \frac{560}{7} = 80$$

Average annual exports (in Rs.crore) of Company Z during the given period

$$= \frac{1}{7} \times (60+90+120+90+60+80+100) = (\frac{600}{7})$$

Therefore, required percentage $= \frac{80}{(600/7)} \times 100]\%$

$=$ approximately 93.33%

3. In which year was the difference between the exports from Companies X and Y the minimum?

    A. 2012

    B. 2013

    C. 2014

    D. 2015

Answer: C

The difference between the exports from the Companies X and Y during the various years are:

In 2011, Rs.(80-30) crores = Rs.50 crores.

In 2012, Rs.(60-40) crores = Rs.20 crores.

In 2013, Rs.(60-40) crores = Rs.20 crores.

In 2014, Rs.(70-60) crores = Rs.10 crores.

In 2015, Rs.(100-80) crores = Rs.20 crores.

In 2016, Rs.(100-50) crores = Rs.50 crores.

In 2017, Rs.(140-120) crores = Rs.20 crores.

Clearly, the difference is minimum in the year 2014.

Pie Chart

The following pie-chart shows the percentage distribution of the expenditure incurred in publishing a book. Study the pie-chart and the answer the questions based on it.

Various Expenditures (in percentage) Incurred in Publishing a Book

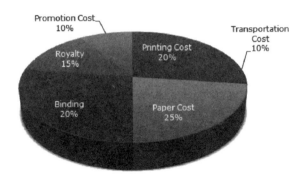

1. If for a certain quantity of books, the publisher has to pay Rs.30,600 as printing cost, then what will be amount of royalty to be paid for these books?

   A. Rs.19,450
   B. Rs.21,200
   C. Rs.22,950
   D. Rs.26,150

Answer: C

Let the amount of Royalty to be paid for these books be Rs.r.

Then, $20:15 = 30600:r \rightarrow r = Rs. \dfrac{(30600 \times 15)}{20} = Rs.22,950$

2. What is the central angle of the sector corresponding to the expenditure incurred on Royalty?

    A.   15°

    B.   24°

    C.   54°

    D.   48°

Answer: C

Central angle corresponding to Royalty = (15% of 360)°

$= (\dfrac{15}{100} \times 360)°$

$= 54°$

3. The price of the book is marked 20% above the C.P. If the marked price of the book is Rs.180, then what is the cost of the paper used in a single copy of the book?

    A.   Rs.36

    B.   Rs.37.50

    C.   Rs.42

    D.   Rs.44.25

Answer: B

Clearly, marked price of the book = 120% of C.P.

Also, cost of paper = 25% of C.P

Let the cost of paper for a single book be Rs.n.

Then, $120:25 = 180:n \rightarrow n = Rs.(\frac{25 \times 180}{120}) = Rs.37.50$

4. If 5500 copies are published and the transportation cost on them amounts to Rs.82500, then what should be the selling price of the book so that the publisher can earn a profit of 25%?

   A. Rs.187.50
   B. Rs.191.50
   C. Rs.175
   D. Rs.180

Answer: A

For the publisher to earn a profit of 25%, S.P. = 125% of C.P.

Also Transportation Cost = 10% of C.P.

Let the S.P. of 5500 books be Rs.x.

Then, $10:125 = 82500:x \rightarrow x = Rs.(\frac{125 \times 82500}{10}) = Rs.1031250$

Therefore, S.P. of one book = $Rs.(\frac{1031250}{5500}) = Rs.187.50$

CHAPTER 20

ESSAY WRITING

引題寫作技巧

關於作文的引題部分，以下給大家推薦幾個作文開頭的寫作模式，希望對大家有所幫助：

1. 敘開型

先概述申論材料的主要內容或反映的主要問題，然後再亮開觀點。這一順序屬於正常順序，也最常用。

2. 引言型

以名言警句、諺語俗語、理論文獻等開頭，然後亮出中心論點。

思維結構：名言警句/諺語俗語/理論文獻+分析→論點

3. 概述型

開頭或概述事例，或概述現象，或援引數據，然後帶出中心論點。

思維結構：事例/現象/數據+分析→論點

4. 點題型

開門見山，或闡釋內涵，或闡釋關係，然後帶出中心論點。

思維結構：內涵/關係+分析→論點

5. 設疑型

以對問題或現象的建設性提問開頭，或自問自答，或問而不答，帶出中心論點。

思維結構：設問/疑問＋分析一論點

其他尚有設喻型，即打個比方來引出論點，以及設問型，即以設問引出論點。

注意：寫作題主要測試考生的語文及組織能力。不少考生均擔心其文章會因「政治不正確」而落選，但其實無立場才是近年不少考生的真正死因。很多考生筆試只着重正反論點，但除了展示對題目的認識外，如何將個人的觀點融入題目，對事件的建議同樣重要，考生應該表現得更有主見。

【例題】

《論語》有云：「不學禮，無以立。」請以這句話為中心議題，結合社會現實，自擬題目，寫一篇文章。

1. 轉折遞進式

思維結構：好現象描述＋問題描述＋分析問題＋亮明觀點

改革開放以來，內地經濟發展取得舉世矚目的成就，可以說，中國已經開始在國際舞台上展現著大國的姿態。（好現象）然而，當前卻暴露了中國人的一系列質素問題，不得不引起我們的重視，文明缺失現象時有發生，傳媒爆出內地人的行為也屢見不鮮。（問題）這無疑會對中國樹立大國意識、彰顯大國形象造成強烈衝擊。（分析）大國崛起，不僅要有物質文明作為基礎，更要有文化軟實力作支持，提高公民質素影響深遠，中國人必須學禮才能更好地樹立大國風範。（帶出觀點）

2. 概括式

思維結構：主要問題＋分析原因＋點出觀點

當前，中國的物質文明走得太快，靈魂卻沒有跟上。中國人的一系列文明質素問題令人堪憂，特別是中國旅客在外國一系列行為，

如：公共場合亂扔垃圾、大聲喧嘩、不守秩序，亂刻亂畫，不僅是自身文明質素低下的表現，同時也是對自己國家的一種抹黑。（主要問題）出現這種情況，除了反映出人民自身文化修養不夠，更反映出中國在精神文明建設方面存在缺失。（分析原因）為此，必須要引導公民學禮提升文化修養，更好地彰顯大國風範。（帶出觀點）

3. 案例分析式

思維結構：案例+案例分析+亮明觀點

中國自古是禮儀之邦，在五千年的文明傳承中，形成了獨一無二的模範。航海家達伽馬在到達非洲時樹起旗幟以標示葡萄牙皇室的主權，而與此形成鮮明對比的是早在100多年前，中國航海家鄭和在到達非洲時，卻樹立了一座豐碑，拒絕侵略，傳遞友好。（案例）鄭和的事跡，不僅體現了中國「以和為貴」的傳統文化，更彰顯了我中華氣度。（案例分析）時至今日，我相信中國人依然需要通過學禮去提升大國意識，樹立大國風範。（亮明觀點）

4. 排比式

思維結構：排比+原因+亮明觀點

學禮，是一個人提升文化修養的必要途徑；學禮，是一個民族長期屹

立不倒的重要支撐；學禮，是一個國家實現繁榮富強的根本保障。

也可以反著説，一個不學禮的人，必然是一個缺乏修養、道德水平的人；一個不學禮的民族，必然是一個風雨飄搖、曇花一現的民族；一個不學禮的國家，必然是一個故步自封、走向毀滅的的國家。（排比）

學禮，對於每個人、每個民族和每個國家都至關重要。然而，當前中國卻出現了一系列國民質素低下、文化修養缺失的現像，傳統禮儀正遭受著嚴峻挑戰。究其原因，在於對學禮重視不足。（原因）因此，必須加強學禮，彰顯大國風範。（帶出觀點）

5. 時間式

思維結構：回首過去＋把握現在＋展望未來

中國自古是禮儀之邦，唐朝時期，經濟文化處於世界先進地位，在中、日密切的經濟文化交流中，開放包容，以禮彰顯了大國風範。（過去）孔子曾説：「人無禮則不生，事無禮則不成，國無禮則不守。」當前中國經濟發展取得舉世矚目的成就，而中國人的一系列文明質素問題卻令人堪憂。（現在）展示大國姿態，不僅需要物質文明做基礎，更需要文化軟實力做支撐。實現「中國夢」，需要弘揚禮文化，只有學禮，才能樹立大國風範。（未來）

結尾寫作技巧

好的文章結尾的要求基本上是——言簡意賅，蒼勁有力，總括全文，適當升華。同樣，一篇優秀的申論文章，離不開一個精彩的結尾。

1. 結尾的標準

申論寫作文章的結尾需做到簡潔、總結全文和升華主題。

a. 簡潔。這點與開頭相似，結尾部分要乾淨利落，古人形容文章「鳳頭豬肚豹尾」，豹的尾巴靈動迅捷，比喻文章的結尾部分要通過簡短的段落，讓閱讀者再次了解作者論述的主題和主要觀點。

b. 總結全文。結尾同樣也要體現立意，其最為基本的要求就是在結尾部分能回扣主題，總結全文內容。這樣能夠使文章前後呼應，內容緊湊。

總結全文一般有兩個方法：一方面，要再次展示總論點，起到點題的作用；另一方面，要概括各個分論點，相當於對文章主要內容的總結。

c. 升華主題。這是結尾的更高的要求，在達到形式上簡潔和內容上總結全文的要求的基礎上，適當升華主題。但是要注意，升華不等於拔高文章，也不等於叫口號，這樣會使文章結尾脫離論述的主題，弄巧成拙。

2. 寫作技巧

結尾標準從形式與內容上規範了申論文章的結尾，考生們在寫結尾時，應從掌握基礎寫法開始，儘量做到篇幅短小，內容紮實。

a. 再現總論點

最後一段發揮的作用就是收束全文，但仍需要兼顧點明文章主旨，最簡單的方式就是複述總論點，但注意不是照抄總論點，而是用換一種表達方式展現出總論點。

以下，我們用「培育社會道德要落小、落細、落實」主題為例：

【範例】行遠，必先修其近；登高，必先修其低。近不修，無以行遠路；低不修，無以登高山。社會是我們共同的家庭，每位家庭成員都需要從一點一滴做起，踐行道德規範、提高精神文明，養成良好習慣、培養高尚品質，共同建立我們的精神家園。

【解析】上述結尾部分前半部分，引用《中庸》中的一句警世名言：「行遠，必先修其近；登高，必先修其低。近不修，無以行遠路；低不修，無以登高山。」意思是説：要想走很遠的路，到達遠

大的目標，就必須從近處開始；要想登上高山之巔，極目遠眺，一覽眾山小，就必須從山腳起步。接近「行遠自邇，登高自卑」，表示凡事要從細微處做起。

後半部分闡釋個人與社會的關係，並講到要培育良好的社會道德就得每個人自覺地「踐行道德規範、提高精神文明，養成良好習慣、培養高尚品質」，點明文章「培育社會道德要落小、落細、落實」的中心論點。

b. 概括各個分論點

總結全文的另一種方法就是概括前面文章主體部分的內容，由於文章主體部分是圍繞分論點展開進行論述的，因此考生在結尾的部分，也可以再次將各個分論點提煉出來，形成總結，起到收束文章內容的作用。

接著我們用「促進科技成果轉化」的主題為例，分論點是「提升高校的科技研發水平」、「加強企業與高等院校技術研發的交流互動」和「政府提出積極的激勵政策」等。

【範例】高等院校科技成果轉化除了需要科技成果自身過硬之外，還需要政府職能的轉變和企業與高校的通力合作。當前，促進高校科技成果轉化的良好政策激勵環境正在形成。「好風憑藉力，送我上青天。」我們可以這樣說，三方合力共促科技成果轉移轉化的 「春天」來了！

【解析】範例所示第一句，通過並聯連詞將三個分論點的內容串聯起來，總結了促進科技成果轉化的三方力量。並通過引述背景，展示政府政策激勵的作用，對未來科技成果轉化作出了展望，點明了三個措施共同發力定能實現科技成果高效轉化的美好前景。

人們說：「好的開頭，是文章成功的一半」。其實，好的結尾，又何嘗不是優秀範文的保障呢？結尾，是文章的畫龍點睛之筆，也是體現文章作者對整個文章的規劃與構思，無論是複述總論點還是概括分論點，其實都是點明文章的立意，並適度升華主旨。

CHAPTER 3

TEST PAPER

Instruction:

Format: The examination consists of two papers: Paper 1 and 2.

The format and content of the two papers are as follow:

Paper 1: Language & Aptitude Test (MC questions)
Time Limit: 1 hour 30 minutes

This paper consists of the following question types:

(i) Language Test

- Comprehension

- Cloze

- Error Identification

- 閱讀理解（題式以中文提問）

- 詞運用（題式以中文提問）

- 字詞辨析（題式以中文提問）

- 句子辨析（題式以中文提問）

(ii) Aptitude Test

- Verbal Reasoning (English)

- Data Sufficiency

- Interpretation of Tables and Graphs

Paper 2: Essay
Time Limit: 2 hours

Candidates are required to write an English essay on a given topic in not less then 500 words, and a Chinese essay on a given topic in not less than 600 words.

Paper 1

(Time Limit: 1 hour 30 minutes)

A. Comprehension

Read the following passage and answer questions 1-10. For each question, choose the best answer from the given choices.

A fact that draws our attention is that, according to his position in life, an extravagant man is either admired or loathed. A successful business man does nothing to increase his popularity by being prudent with his money. A person who is wealthy is expected to lead a luxurious life and to be lavish with his hospitality. If he is not so, he is considered mean, and his reputation in business may even suffer in consequence. The paradox remains that he had not been careful with his money in the first place; he would never have achieved his present wealth.

Among the low income group, a different set of values exists. The young clerk, who makes his wife a present of a new dress when he has not paid his house rent, is condemned as extravagant. Carefulness with money to the point of meanness is applauded as a virtue. Nothing in his life is considered more worthy than paying his bills. The ideal wife for such a man separates her housekeeping money into joyless little piles – so much for rent, for food, for the children's shoes, she is able to face the milkman with equanimity every, month satisfied with her economizing ways , and never knows the guilt of buying something she can't really afford .

As for myself, I fall neither of these categories. If I have money to spare I can be extravagant, but when, as is usually the case, I am hard up and then I am the meanest man imaginable.

1. Which of the following would be the most appropriate title for the passage:

   A. Being extravagant is always condemnable.
   B. The cause of poverty is extravagance.
   C. Extravagance is a part of the rich as well as of the poor.
   D. Stingy habits of the poor.

2. According to the passage the person, who is a successful businessman and wealthy

   A. Is expected to have lavish lifestyle.
   B. Should not bother about popularity.
   C. Is more popular if he appears to be wasting away his time.
   D. Must be extravagant before achieving success.

3. The phrase "lavish with his hospitality" in the third sentence of the first paragraph means

   A. Thoughtful in spending only on guests and strangers.
   B. Unconcerned in treating his friends and relatives.

C. Stinginess in dealing with his relatives.

D. Extravagance in entertaining guest.

4. The word "paradox" in the last sentence of the first para-graph means

A. Statement based on the popular opinion

B. a statement that seems self-contradictory but in reality ex-presses a possible truth.

C. Statement based on facts

D. A word that brings out the hidden meaning

5. What is the meaning of the word "equanimity"?

A. Calmness

B. Discomposure

C. Equivocal

D. Dubious

B. Cloze Test

Read the following passage and answer questions. For each question, choose the best answer from the given choices.

To others and themselves the British have a reputation for being conservative - not in the narrow political sense, but in the sense of adherence to accepted ideas and unwilling to question them. The reputation comes partly from their ___(1)___. For 900 ___(2)___ they have suffered ___(3)___ invasion nor revolution (except in 1649 ___(4)___ 1688) nor disastrous defeat in ___(5)___ . Their monarchy survives ___(6)___ serious question. Under its normal ___(7)___ the political arrangements have been ___(8)___ stable that, except for the ___(9)___ interruptions in the seventeenth ___(10)___ , they have been adopted throughout ___(11)___ centuries to meet changing needs without violent ___(12)___. Britain, in 1978, was ___(13)___ in managing without ___(14)___ written constitution; some fragmentary definitions of 1688 still ___(15)___. There had been ___(16)___ quarrels, social and economic as well ___(17)___ political, but the quarrels had been ___(18)___, usually ___(19)___ compromise. The underlying ___(20)___ had not been broken.

(1) A. language   B. future   C. history   D. literature

(2) A. centuries   B. years   C. minutes   D. seconds

(3) A. neither   B. as   C. or   D. either

(4) A. too   B. but   C. or   D. and

(5) A. home   B. study   C. peace   D. war

(6) A. and   B. without   C. with   D. neither

(7) A. name   B. people   C. leadership   D. enemy

(8) A. so   B. very   C. too   D. such

(9) A. two   B. one   C. no   D. couple

(10) A. month   B. day   C. year   D. century

(11) A. a   B. any   C. the   D. few

(12) A. changes   B. change   C. altered   D. changed

(13) A. common   B. popular   C. unique   D. angry

(14) A. two   B. a   C. some   D. their

(15) A. survive   B. surviving   C. survives   D. survived

(16) A. bitter   B. happy   C. pleasant   D. unhappy

(17) A. for   B. as   C. to   D. at

(18) A. settling   B. settles   C. settle   D. settled

(19) A. for   B. at   C. to   D. by

(20) A. continue   B. continuity   C. continuous   D. continued

C. Error Identification

Each of the sentence below may contain a language error. Identify the part (underlined and lettered) that contains the error or choose "(E) No error" where the sentence does not contain an error.

1. Do you mind to swap offices with me this afternoon? We've got a departmental meeting, but we can't use the usual meeting room because it's being renovated. You're room's a bit roomier than mine is. Hope I'm not being too awkward!

   A. to swap
   B. being renovated
   C. roomier
   D. being too awkward
   E. no error

2. Ice cream sales vary according to the time of year. In summer, sales boom as everyone is in trying to keep cool, and people are in the mood for ice cream. Then sales plummet in the autumn and remain low during the winter, although they raise a little over the festive season when a lot of people are eating.

   A. boom
   B. plummet
   C. remain low
   D. raise
   E. no error

3. We need to book a temp to cover Rebecca while she's away on annual leave. Would you mind getting in touch to the temping agency? The dates are 14th-8th July.

A. away
B. on
C. in
D. to
E. no error

4. On your resumé, you should write a brief history of your education and training, mentioning your qualifications and the grades you obtained. Also include a work history. Include a section about your personal interests, and don't forget to write the contact information for two referrals.

A. history
B. qualifications
C. grades
D. referrals
E. no error

D. 閱讀理解

1. 政府向社會組織購買服務，是通行於國際社會的一種新型的公共服務供給方式。其具體做法是政府引入市場機制，把直接向社會公眾提供的一部分公共服務和社會管理事項，按照一定的方式與程序，交由具備條件的社會組織承擔，並根據服務數量和質量向社會組織支付費用。

   根據上述定義，下列屬於政府向社會組織購買服務做法的是：

   A. 某街道面向轄區招募志願者，以協助當地司法機關承擔社區內服刑人員的社區矯正服務工作。

   B. 某市交管部門為方便市民出行，租用電訊公司的短訊平台定期給市民發送提醒短訊。

   C. 某市教委建立特級教師在線回答中小學生學習問題的公益網站受到了學生和家長的好評，一家教育機構出資贊助擴大這個網站的規模，以幫助更多的學生。

   D. 國家為保護農民利益和種糧積極性，委托中儲糧總公司作為托市收購政策的執行主體，組織國有糧庫對稻農實行敞開收購。

2. 「技術外溢」是指外商投資、跨國貿易等對東道國相關產業或企業的產品開發技術、生產技術、管理技術、營銷技術等產生的提升效應。包括平行外溢和垂直外溢。對當地競爭企業的技術創新的示範、刺激與推動，稱為平行外溢；對當地上下游關聯企業的技術進步的示範、援助與帶動，稱為垂直外溢。

   根據上述定義，下列不屬於「技術外溢」的一項是：

A. 某跨國公司按照技術規範對上游產品生產企業提供圖紙，並派出技術人員對上游企業的生產加以指導。

B. 某外資集團給當地帶來了新技術、新理念，為同領域的本土企業提供了樣本示範，很多本土企業通過模仿改進實現了自主創新。

C. 國內兩家互聯網巨頭互相挖對方的技術和管理人才，競爭日益激烈，也推動雙方的產品越做越精細。

D. 某跨國公司進入本土行業後，依靠先進技術迅速佔據市場，本土企業為保住自己的市場份額不斷改進技術和工藝，改善產品結構，從而激發了自主開發能力。

3. 「灰色經濟」又稱地下經濟，是國民經濟體的一部分。灰色經濟包括所有創造價值的經濟活動，但是它們沒有被列入國民經濟總決算中，即沒有被包括在正式公布的國民生產總值中。灰色經濟又可以分成兩個部分：一種是自給自足經濟，指的是生產的目的不是為了交換，而是為了直接滿足本經濟單位或生產者個人的需要；還有一種地下經濟，它雖然應當被納入決算，但因為各種原因（比如隱瞞）而未被列入。

根據上述定義，以下不屬於自給自足經濟的是哪項？

A.張先生在自己的後院開墾了一片菜地，種些蔬菜供自家人食用。

B. 小明家養了兩頭豬，為了過年時準備全豬宴

C. 某村農閒時幾名婦女做些手工活供自己家人穿用

D. 某人從香港逃稅帶回幾部蘋果手機

E. 語句排序

1. 選出下列句子的正確排序：

(1) 通過龍場悟道，陽明先生不但發現了一切人都可以成為聖賢之人的潛在性和可能性，同時還給人們指引了一條成聖的道路

(2) 他的思想在大多數人眼中雖如陽春白雪，卻對當今社會有著重要的現實意義

(3) 明先生乃一「立功、立德、立言」的大儒聖賢

(4) 即證到「聖人之道，吾性自足，向之求理於事物者誤也」

(5) 他於500年前，在貴州龍場這個天高皇帝遠、蛇蠍滿地的地方，實現了一次驚天動地的「龍場悟道」

A. 3-5-2-4-1
B. 2-1-3-5-4
C. 3-5-4-1-2
D. 3-4-5-1-2

2. 選出下列句子的正確排序：

(1) 這需要內地乳業拿出誠意，利潤血管裡流淌道德血液，加強行業自律，從奶源到生產、銷售諸環節都嚴格把關，不幹搬起石頭砸自己的腳的蠢事

(2) 對於內地國產奶粉來講，價格相對便宜，無疑是競爭優勢籌碼

(3) 職能部門也要提供權威説法與科學檢測數據給國產奶粉正名

(4) 但是，國產奶粉能否抓住羊奶粉頻頻漲價的契機，振興中國乳業，重新取得消費者的信任奪回市場，還有許多事情要做

A. 1-2-3-4

B. 1-3-2-4

C. 2-4-1-3

D. 2-1-3-4

3. 選出下列句子的正確排序：

(1) 認為瑪雅文明和中國古代文明是同一祖先的後代在不同時代、不同地點發展的結果

(2) 其主要根據是《梁書》中關於五世紀時中國僧人慧深飄洋過海到達「扶桑國」的故事，認為「扶桑」即墨西哥

(3) 日前，著名人類學家張光直教授提出了一個「瑪雅──中國文化連續體」的假設

(4) 遺憾的是，這些説法至今還沒有得到考古學的明確證實

(5) 後來又有「殷人東渡説」，是説商朝時的中國人橫渡太平洋將文明帶到了美洲

(6) 長期以來，不少人認為瑪雅文明的源頭是古代的中國文明，最早提出這種觀點的是「扶桑國」説

A. 3-6-1-2-5-4

B. 3-1-4-6-2-5

C. 6-3-2-5-1-4

D. 6-2-5-3-1-4

F. 字詞辨析：錯別字

1. 請選出沒有錯字的句子：

   A. 他喜歡過悠閒的日子，不喜歡忙碌的生活
   B. 練習書法，可以含養自己的身心。
   C. 在無可耐何之下，他承認所有的過錯。
   D. 這個消息傳到學校，人人都感到無比震憾。

2. 請選出沒有錯字的句子：

   A. 身為老師，就應該付起傳道、授業、解惑的責任
   B. 為了做好我的工作，每天都競競業業，不敢懈怠
   C. 我們的士兵，兵驍將勇，足以維護國家安全
   D. 他是出了名的銘頑不靈，想要跟他溝通，談何容易

3. 請選出沒有錯字的句子：

   A.　在各地廟宇舉辦建醮大典時，虔誠的善男信女，往往一擲千金，毫不吝嗇
   B.　不良少年風馳電掣的飛車表演，使得馬路交通癱瘓，路上行人個個提心弔膽
   C.　多少人寅緣富貴，他卻腳踏實地，按部就班，憑自己的努力求取榮譽
   D.　這裡原是四通八達的衛道，但在地震之後，變得滿目瘡痍，慘不忍睹

4. 請選出沒有錯字的句子：

    A. 此等議論，正是推波逐瀾，縱風止燎

    B. 所以說為官做吏的人，千萬不要草管人命，視同兒戲

    C. 韓愈《進學解》自述焚膏繼晷，兀兀窮年，終成一代宗師

    D. 怎當得元兵勢大，宋軍未曾交綏，先自望風披靡

5. 請選出沒有錯字的句子：

    A. 讀書人不應貪圖生活享受，衣取敝寒，食取充腹即可

    B. 新市長力求興利除弊，整頓長久以來為人垢病的積習

    C.結婚喜宴席開數百桌太過侈靡，但完全不宴客似乎又矯枉過正

    D. 只重居弟華美卻不重內在修養的俗人，令人相當鄙示

G. 字詞辨析：簡體字運用

1. 請選出下面繁體字錯誤對應簡化字的選項：

   A. 實（实）
   B. 頭（实）
   C. 氣（气）
   D. 鹵（卤）

2. 請選出下面繁體字錯誤對應簡化字的選項：

   A. 憲（宝）
   B. 骯（肮）
   C. 淚（泪）
   D. 瞭（了）

3. 請選出下面簡化字錯誤對應繁體字的選項：
   A. 拥（擁）
   B. 余（餘）
   C. 齐（齊）
   D. 庆（庆）

4. 請選出下面簡化字錯誤對應繁體字的選項：
   A. 只（隻）
   B. 帘（窮）
   C. 条（條）
   D. 参（參）

H. 句子辨析：語病句

1. 下列各句中，沒有語病的一句是：

A. 中國科學院最近研究發現，喜馬拉雅山冰川退縮，湖泊的面積擴張，冰湖潰決危險性增大，引起了研究者的廣泛關注。

B. 長江中的江豚被譽為「水中大熊貓」，是國家二級保護動物，也是《華盛頓公約》確定的全球瀕危物種之一，再不加以保護，15年後將會滅絕。

C. 專家認為，中國人均飲茶量每天不足10克，加之大部分農藥不溶於水，茶葉中即使有少量的農藥殘留，泡出的茶湯中也會農藥含量極低，對人體健康影響不大。

D. 今年廣東天氣形勢複雜，西江、北江可能出現五年一遇的洪水；省政府要求各地要立足防大汛、搶大險、抗大旱，做到排查在前、排險在前、預警在前，確保群眾的生命財產安全。

2. 下列各句中，沒有語病的一句是：

A. 專家認為，如今衡量一部文學作品的價值大都依據市場銷量，這缺少足夠的權威性與公信力，眼下不少暢銷書本身並不具備多少文學價值，只因為炒作等原因而受歡迎。

B. 從2018年7月1日開始，遊樂場的門票價格由300元調整為500元。這一門票漲價的消息引起了很多市民、遊客、網友們和參觀者的熱議。

C. 目前，文物保護狀況雖然得到明顯改善，但文物保護工作中仍存在認識不足、文物安全形勢嚴峻、經費人才不足，其中文物保護與經濟建設的矛盾比較突出。

D. 之所以有很多作者選擇獨立動畫創作且樂此不疲，根本原因就是因為這種表達方式很有趣，個人色彩很濃厚，能突破很多既定樣式的牢籠，是一種真正「創造」出來的文化產品。

3. 下列各句中，沒有語病的一句是：

A. 文化遺產是一種不可再生的珍貴資源。當前，隨着經濟全球化趨勢和現代化步伐，世界文化的多樣性面臨着一體化的趨勢，文化遺產的保護形勢非常嚴峻。

B. 不知不覺間，「微時代」已經來到了我們面前。微博取代博客成為網絡上最主流的信息分享平台，微電影也逐漸成為年輕觀眾躲避影院爛片「轟炸」的首選。

C. 平民文化的重新崛起，不僅會對中國整個社會文化的建設大有裨益，更會對中國電影產業的發展起到巨大推動作用。

D. 誠信教育已成為一個國家公民道德建設的重要內容，因為不僅誠信關係到國家的整體形象，而且體現了國民的基本道德質素。

4. 下列各句中，沒有語病的一句是：

A. 可口可樂飲料有限公司近日確認，在實施管道改造時，由於操作失誤，導致含微量餘氯的生產輔助用水進入到飲料生產用水中。

B. 出版業當然要講究裝幀藝術，講究宣傳造勢和市場營銷，但要想真正贏得讀者、贏得市場，最終還是取決於內容是否具有吸引力和感染力。

C. 網絡謠言的特點就在於傳播的迅猛和來源的不確定。面對鋪天蓋地的謠言，人們往往容易忽視最基本的事實。

D. 這所創建於上世紀20年代初期的商學院是這座濱海城市的唯一的一所大學，這所大學一直對孩子們充滿了神秘感。

I. 句子辨析:邏輯錯誤

1. 美國的一位動物學家在黑猩猩的籠子前放了一面大鏡子,觀察他們的反應。它們能夠從鏡子中認出自己,經常久久地對著鏡子尋找自己身上平時看不到的部位。作為黑猩猩近親的大猩猩卻不具備這種能力。

   由此可以推出:
   A. 黑猩猩在某些方面的能力更高於大猩猩
   B. 黑猩猩和大猩猩雖然親緣較近,但還是黑猩猩的智力更勝一籌
   C. 黑猩猩的這種能力和它所生活的環境有關
   D. 動物界還有其它動物具備這種能力

2. 已知:(1)如果甲和乙是肇事者,丙就不是肇事者;(2)如果丁是肇事者,那麼乙就是肇事者;(3)甲和丙都是肇事者。

   由此推出:

   A. 乙和丁都是肇事者
   B. 乙和丁都不是肇事者
   C. 乙是肇事者,丁不是肇事者
   D. 乙不是肇事者,丁是肇事者

3. 如果你的筆記型電腦是1999年以後製造的,那麼它就帶有調制解調器。

   上述斷定可由以下哪個選項得出?

A. 只有1999年以後製造的筆記型電腦才帶有調制解調器

B. 所有1999年以後製造的筆記型電腦都帶有調制解調器

C. 有些1999年以前製造的筆記型電腦也帶有調制解調器

D. 筆記型的調制解調器技術是在1999年以後才發展起來的

4. 物理學家霍金認為：「地球生命被某一災難消滅的危險正以前所未有的速度與日俱增突如其來的全球變暖、核戰爭、基因病毒或其他危險。」霍金在此基礎上認為地球越來越不適合人類居住，並提出「太空移民」的觀點。

以下最能質疑霍金觀點的一項是：

A. 絕大多數星球的環境不適合人類生存

B. 人類面臨的危險通過人類的共同努力能夠得到解決

C. 地球處於其生命年齡的幼年時期

D. 地球所擁有的適合居住的表面區域太小，抵禦毀滅性災難的能力差

J. Verbal Reasoning

Passage 1

If society seems obsessed with youth, it is at least partly because companies are. Like it or not, the young increasingly pick the styles and brands that trickle up to the rest of the population. Nike, Abercrombie & Fitch and Timberland first found success with the young, and when that clientele tired of them the companies felt the loss deeply. Now that adults are no longer necessarily expected to act and look grown-up, parents and children can be found listening to exactly the same music, playing the same computer games, watching the same TV programs, and wearing the same brands of clothes and shoes.

1. An adult's style can sometimes be similar to that of a child's.

2. The profits of Timberland are not affected by young customers.

3. Adults wear the same shoes as children because they want to look younger.

Passage 2

Television is changing as it goes digital. The result will not only be better-quality pictures and sound but also personal TV, with viewers able to tailor the programs they watch and even interact with them. How much money this will make for program producers or broadcasters, whoever they may be, is not so clear.

Cable, satellite and terrestrial television broadcasters are upgrading their equipment to provide higher quality digital services. Rupert Murdoch's News Corporation will become the first company in the world to migrate an entire national TV system over to digital when it turns off its old analogue version of its British satellite service, BSkyB.

4. Rupert Murdoch is associated with BSkyB.

5. The only change from traditional analogue services to digital services will be the picture quality.

6. Television broadcasters are upgrading their equipment because they will make more money from digital TV.

Passage 3

Brand equity has become a key asset in the world of competitive business. Indeed, some brands are now worth more than companies. Large corporations themselves are widely distrusted, whereas strangely, brands have the opposite effect on people. Brands are used to humanize corporations by appropriating characteristics such as courage, honesty, friendliness and fun. An example is Dove soap, where a dove represents white, cleanliness and peace. Volkswagen like to give the impression through their advertising that they are a reliable, clever, technical product. In a sense, rather than the product itself, the image

7. Brands have always been an important asset to a company.

8. Many people distrust large corporations.

9. Dove soap chose a dove for their brand to give a sense of cleanliness and peace.

Passage 4

The first problem with financial statements is that they are in the past; however detailed, they provide just a snap-shot of the business at one moment in time. There is also a lack of detail in financial statements, giving little use in the running of a business. Financial statements are provided for legal reasons to meet with accounting regulations and are used mainly by City analysts who compute share prices and give guidance to shareholdeRs.Accounts often have hidden information and may also be inconsistent; it is difficult to compare different companies' accounts, despite there being standards, as there is much leeway in the standards.

10. Financial statements are useful for businesses to understand their financial activities.

11. Companies create financial statements in order to comply with their legal obligations.

12. If account reporting standards were tightened, it would be easier to compare the performance of different companies.

K. Data Sufficieny

1. If a and b are both positive, what percent of b is a?

   (1) a = 3/11

   (2) b/a = 20

   A. Statement (1) alone is sufficient, but statement (2) alone is not sufficient to answer the question.
   B. Statement (2) alone is sufficient, but statement (1) alone is not sufficient to answer the question.
   C. Both statements taken together are sufficient to answer the question, but neither statement alone is sufficient.
   D. Each statement alone is sufficient.
   E. Statements (1) and (2) together are not sufficient, and additional data is needed to answer the question.

2. A wheel of radius 2 meters is turning at a constant speed. How many revolutions does it make in time T?

   (1) T = 20 minutes

   (2) The speed at which a point on the circumference of the wheel is moving is 3 meters per minute.

   A. Statement (1) alone is sufficient, but statement (2) alone is not sufficient to answer the question.
   B. Statement (2) alone is sufficient, but statement (1) alone is not sufficient to answer the question.

C. Both statements taken together are sufficient to answer the question, but neither statement alone is sufficient.

D. Each statement alone is sufficient.

E. Statements (1) and (2) together are not sufficient, and additional data is needed to answer the question.

3. Is $x > 0$?

(1) $-2x < 0$

(2) $x^3 > 0$

A. Statement (1) alone is sufficient, but statement (2) alone is not sufficient to answer the question.

B. Statement (2) alone is sufficient, but statement (1) alone is not sufficient to answer the question.

C. Both statements taken together are sufficient to answer the question, but neither statement alone is sufficient.

D. Each statement alone is sufficient.

E. Statements (1) and (2) together are not sufficient, and additional data is needed to answer the question.

4. A certain straight corridor has four doors, A, B, C and D (in that order) leading off from the same side. How far apart are doors B and C?

(1) The distance between doors B and D is 10 meters.

(2) The distance between A and C is 12 meters.

A. Statement (1) alone is sufficient, but statement (2) alone is not sufficient to answer the question.

B. Statement (2) alone is sufficient, but statement (1) alone is not sufficient to answer the question.

C. Both statements taken together are sufficient to answer the question, but neither statement alone is sufficient.

D. Each statement alone is sufficient.

E. Statements (1) and (2) together are not sufficient, and additional data is needed to answer the question.

5. Two socks are to be picked at random from a drawer containing only black and white socks. What is the probability that both are white?

(1) The probability of the first sock being black is 1/3.

(2) There are 24 white socks in the drawer.

A. Statement (1) alone is sufficient, but statement (2) alone is not sufficient to answer the question.

B. Statement (2) alone is sufficient, but statement (1) alone is not sufficient to answer the question.

C. Both statements taken together are sufficient to answer the question, but neither statement alone is sufficient.

D. Each statement alone is sufficient.

E. Statements (1) and (2) together are not sufficient, and additional data is needed to answer the question.

Answer:

A. Comprehension

1. C This is the most appropriate title as the author sheds light on the life of a rich person in the first paragraph and discusses the life of a person who belongs to lower strata of the society in the second paragraph.

2. A It is given in the fourth line of the 1st paragraph: "If he ...consequence."

3. D The meaning of the word "hospitality" is the quality or disposition of receiving and treating guests and strangers in a warm, friendly, generous way which justifies that option d is the correct answer choice.

4. B The meaning of the word "paradox" is given in option B and it is also explained in the last sentence of the first paragraph.

5. A The word "equanimity" means mental or emotional stability or composure, especially under tension or strain as can further be understood as in this example "Raj was a man of great equanimity, even when talking about his own death."

B. Cloze Test

1. C。由於下文緊接著敘述到過去900年的有關情況。故應選 history，才能與下文銜接。

2. B。根據句子意思判斷。

3. A。應選neither才能與其後出現的nor構成並列連詞neither...

nor...。

4. **D**。根據句子意思判斷。

5. **D**。需選一名詞與前面的介詞in構成介詞短語，且這個短語在語意上要與前面的disastrous defeat相吻合。故選D可以滿足這個條件。

6. **B**。「without serious question」意為「沒有嚴重的問題」。

7. **C**。「在……領導下」的表達方式是under the...leadership.

8. **A**。「so...that...」構成固定短語，意為「這麼……以致於……」

9. **A**。上面提到1649和1688兩次革命，所以A項正確。

10. **D**。根據句子意思判斷。

11. **C**。the centuries特指那些穩定的世紀。

12. **A**。應從A或B中選一名詞作介詞without的賓語。將A和B加以比較：A. changes作可數名詞用，意為「變化」，符合題意。B. change一詞系不可數名詞，意為「零錢，找頭」。

13. **C**。根據上下文意思，應選unique「獨一無二的；唯一的」。

14. **B**。沒有一項成文憲法，英文字母「a」為一項的意思。

15. **D**。應選survived才與上下文的時態一致。

16. **A**。bitter意為「尖銳的，厲害的」。

17. **B**。as well as為一並列連詞。

18. **D**。settled符合語法要求。

19. **D**。by compromise意為「通過折衷、妥協的辦法」。如：We should settle our differences by compromise.（我們應採取折衷的辦法來解決我們之間的分歧）

20. B。四個選項中,只有名詞continuity正確。

C. Error Identification

1. **A** Use a gerund (-ing) after "mind"

2. **D** "Raise" should be "rise". You raise something – it is a transitive verb. Rise is an intransitive verb.

3. **D** To "get in touch with someone" means to contact someone.

4. **D** Someone who recommends you to a job or training course is called a "referee".

D. 閱讀理解

1. **B**。政府向社會組織購買服務定義的要點為:(i)政府; (ii)直接向社會公眾提供的一部分公共服務或社會管理事 項;(iii)支付費用。由於A和C兩項不存在購買行為,不符 合定義要點(iii);D項不符合要點(ii)。B項符合定義,故 本題答案為B。

2. **C**。技術外溢定義的關鍵信息是:外商投資、跨國貿易對東道 國相關產業或企業的提升效應。C項並不涉及外商投資或跨國 貿易,不符合定義的關鍵信息;A項屬於垂直外溢,B、D兩 項屬於平行外溢,均屬技術外溢的範疇。故選C。

3. **D**。由題幹定義可知,自給自足經濟的定義要點是:(i)生 產的目的不是為了交換;(ii)為了直接滿足本經濟單位或

生產者個人需要。通觀4個選項，D選項「逃稅」不符合自給自足經濟的定義要點，不屬於自給白足經濟，其他各項均符合，故正確選項為D。

E. 語句排序

1. **C**。首先根據首句排除，句（2）以「他」開頭，指代不明，故不能做首句，根據寫作思路（5）提出龍場悟道，（4）是龍場悟道的同義轉述，（1）是通過農場悟道發現的道理，故正確順序應該是（5）、（4）、（1），答案選C。

2. **C**。句（1）含有不確定指代對像的指代詞，不能做首句，排除選項A、B。句（4）內地國產奶粉能否重新取得消費者信任，奪回市場，有許多事情要做，而（1）就是國產奶粉需要做的事情，需要加強行業自律，需要嚴格把關。因此（4）、（1），排除D。故本題選C。

3. **D**。句（2）和（6）都提到了「扶桑國」，應捆綁在一起，另外（2）是對（6）「扶桑國」説的解釋，順序為（6）、（2），排除A、C。本題有表示時間順序的詞語，（3）日前，（6）長期以來，順序為先（6）再（3），排除B。故本題選D。

F. 字詞辨析：錯別字

1. **A** 正確寫法為：B. 涵養　C. 無可奈何　D. 無比震撼
2. **C** 正確寫法為：A. 負起　B. 兢兢業業　D. 冥頑不靈

3. C 正確寫法為：A. 建醮大典　B. 提心吊膽　D. 滿目瘡痍

4. C 正確寫法為：A. 推波助瀾　B. 草菅人命　C. 望風披靡

5. C 正確寫法為：A. 衣取蔽寒　B. 詬病　C. 鄙視

G. 字詞辨析：簡體字運用

1. B

2. A

3. D

4. B

H. 句子辨析：語病句

1. B。 觀乎A項，句式雜糅，研究發現的是後文的三種情況，而「引起廣泛關注」的主語是前文中的三種情況，此「三種情況」既已經作為「研究發現」的賓語，則不可再作引起的主語，可去掉最後一句；C項，關聯詞語使用不當，本句主語為「茶湯」，前一個分句的主語是「茶葉」，將「即使」調至「茶葉」的前面，使之成為讓步狀語；D項，邏輯順序錯誤，「排查在前、排險在前、預警在前」應該改為「預警在前、排查在前、排險在前」。

2. A。 觀乎B項，概念並列不當，把「和參觀者」刪掉；C項，成分殘缺，在「人才不足」後面加「等問題」；D項，句式雜糅，「根本原因就是因為」改成「根本原因就是」或「是因為」。

3. **B**。觀乎A項，成分殘缺，在「步伐」後加「的加快」；C
   項，語序不當，把「不僅」與「更」後面的內容互換；D項，
   語序不當。主語一致時，關聯詞在主語後，應為「誠信不
   僅……而且……」

4. **C**。觀乎A項，結構混亂，「由於……導致……」，無主；B
   項，邏輯混亂，一面對兩面；D項，主客體倒置。

I. 句子辨析：邏輯錯誤

1. **A**。第一步：抓住每句話中的對象

   第二句說明黑猩猩能夠從鏡子中認出自己並對著鏡子尋找自
   己身上平時看不到的部位，第三句說明黑猩猩的近親大猩猩
   不具備第二句中提到的黑猩猩的能力。

   第二步：判斷整體關係

   第一步中的兩句話說明黑猩猩在「照鏡子」這一方面要比大
   猩猩能力強。

   第三步：逐一判斷選項的作用

   A中黑猩猩在某些方面的能力更高於大猩猩，正是對第二步中
   結論的進一步總結，因此A項正確；B項單從「照鏡子」這個
   行為就判斷黑猩猩的智力比大猩猩更勝一籌是不科學的，智
   力需要的是多方面的綜合考察；C項題幹中比沒有涉及「黑猩
   猩所生活的環境」，無從得知黑猩猩的這種能力和它所生活
   的環境是否有關；D項題幹沒有涉及，無法判斷。綜上，故正
   確答案為A。

2. **B**

3. **B**。本題屬於翻譯推理型，主要考查複句的翻譯。題幹可以翻譯為「1999年以後→帶調制解調器」。其中A選項可翻譯為「帶調制解調器→1999年以後」；B選項可翻譯為「1999年以後→帶調制解調器」；C選項可翻譯為「有些1999年以前→帶調制解調器」，該選項可能正確，但是不能由其得到題幹中的斷定；D選項可翻譯為「帶調制解調器→1999年以後」。所以選擇B選項。

4. **A**。第一步：找出論點和論據

論點：霍金認為地球越來越不適合人類居住，並提出「太空移民」的觀點。

論據：霍金認為地球生命被某一災難消滅的危險正以前所未有的速度與日俱增突如其來的全球變暖、核戰爭、基因病毒或其他危險。

第二步：逐一判斷選項

A項否定了「太空移民」的觀點，直接削弱了論點。B項削弱了論據，指出通過人類的共同努力能夠得到解決人類所面臨的危機。C項「地球處於其生命年齡的幼年時期」與題幹無關，屬於無關項。D項加強了論點。A項削弱論點強於B項削弱論據，故正確答案為A。

J. Verbal Reasoning

1. Answer: True

The sentence that proves this is "parents and children can be found listening to exactly the same music, playing the same videogames…and wearing the same brands of clothes and shoes [as the young]".

2. Answer: False

The passage says that when "when that clientele [the young] tired of them the companies felt the loss deeply". Meaning that Timberland's profits are in fact closely affected by the purchasing habits of the young.

3. Answer: Can't tell

The passage does not say why adults and children sometimes end up with the same styles. The passage implies that by listening to the same music and wearing the same clothes as young people, adults are not considered grown-up. But the passage does not say why some adults adopt this style (for example it could be price, comfort, or any number of reasons other than wanting to look younger, but we are not told.)

4. Answer: True

This is stated it he passage where it says "Rupert Murdoch's News Corporation… BSkyB", meaning he is therefore associated.

5. Answer: False

The passage says of switching from analogue to digital: "The result will be better quality pictures and sound but also personal TV, with viewers able to tailor the programs they watch…".

6. Answer: Can't tell

This is quite obvious since the passage explicitly says in the last sentence of the first paragraph that it is not clear how much money broadcasters will make from upgrading.

7. Answer: Can't tell

The passage says "Brand equity has become a key asset in…" implying it has not always been. However "important" is distinct from key, and the passage does not discuss if brand equity was considered important before being considered key.

8. Answer: True

The passage says "Large corporations themselves are widely distrusted". Widely implies an opinion held by many people, and the opinion held would be one of distrust.

9. Answer: Can't tell

Whilst the passage does say that the dove gives an impression of "white, cleanliness and peace", and therefore is a successful brand, the passage does not say how or why the company chose the brand in the first place.

10. Answer: False

The passage says that financial statements have "a lack of detail in financial statements, giving little use in the running of a business", which is supported by other critical statements such as 'the first problem with financial statements is that they are in the past" means they cannot be considered useful for businesses to understand their financial activities.

11. Answer: True

This is stated here: "Financial statements are provided for legal reasons to meet with accounting regulations".

12. Answer: Can't tell

The last sentence in the passage says, "it is difficult to compare different companies' accounts, despite there being standards, as there is much leeway in the standards". Though this implies that if there were less leeway in the standards, it would be easier to compare different companies' accounts, this is only an inference. Thus, it can only be asserted as a probability rather than a certainty. Since the passage does not expressly say, we require more information. Therefore, it is cannot say.

K. Data Sufficieny

1. Answer: B

Statement (1) tells us nothing about b and so the answer cannot be A or D. To find what percent a is of b we need to solve the expression a/b x 100.

Statement (2) allows us to do just that: a/b = 1/20. No need to go any further; the answer is B.

2. Answer: C

To find the number of revolutions we need to know the rate of turning and the time duration.

Statement (1) gives us only the time, and so the answer cannot be A or D.

Statement (2) tells us the rate at which a point on the circumference is moving, which, since we know the dimensions of the wheel, is sufficient to determine the number of rotations per minute. But since we do not know the time, B cannot be correct. But putting statements (1) and (2) together we have all we need, so the answer is C.

3. Answer: D

The statement that x is greater than zero means that x is positive. If we multiply a positive number by a negative number the product is negative: this is what we get from statement (1), which thus tells us that x is positive. The answer must be A or D. The cube of a positive number is positive; the cube of a negative number is negative, and so statement (2) tells us that x is positive. And so the answer is D.

4. Answer: E

It is obvious that neither statements (1) or (2) alone can tell you how far apart B and C are, and so the answer must be C or E. To see whether putting both pieces of information together will be adequate, visualize two rods: BD of length 10 units, and AC of length 12 units. Mentally place the rods alongside each other so that C lies between B and D. Now you can mentally slide the rods past each other to see that C can lie anywhere between B and D, and so we cannot fix one value for the length BC, and the answer is E.

5. Answer: C

From statement (1) we know the ratio of black socks to white, but that ratio will change when one sock is taken out. To get the new ratio, and hence the probability that the next sock will also be white, we need to know the number of socks of each type. The answer cannot be A or D. Obviously statement (2) on its own does not get the ratio and so B cannot be correct. But putting the information in both statements together we can solve the problem (24 white socks with a ratio of black to total of 1:3 means that there are 12 black and 24 white socks). The answer is C.

Paper 2: Essay

(Time Limit: 2 hours)

1. 中文作文題目：

替民政事務局局長準備一篇有關紅十字會輸血服務中心揭幕禮的演講辭。

2. 英文作文題目：

致函教育署署長，就改善「學童自殺問題」提出建議。

題目背景：

－學童自殺問題的現況及嚴重性

－解決重要性和迫切性

－學童自殺問題帶來的影響

－建議署方如何改善學童自殺問題

看得喜 放不低

創出喜閱新思維

書名	投考公務員系列 入境事務處筆試全攻略
ISBN	978-988-78874-3-0
定價	HK$128
出版日期	2019年1月
作者	Fong Sir
責任編輯	文化會社投考公務員系統編輯部
版面設計	梁文俊
出版	文化會社有限公司
電郵	editor@culturecross.com
網址	www.culturecross.com
發行	香港聯合書刊物流有限公司
	地址：香港新界大埔汀麗路36號中華商務印刷大廈3樓
	電話：（852）2150 2100
	傳真：（852）2407 3062